High Bias

The University of North Carolina Press

CHAPEL HILL

The Distorted History of the
CASSETTE TAPE

MARC MASTERS

Designed by Lindsay Starr
Set in Miller and Market Pro
by codeMantra

Manufactured in the United States of America

Library of Congress Cataloging-in-Publication Data
Names: Masters, Marc, author.
Title: High bias : the distorted history of the cassette tape / Marc Masters.
Description: Chapel Hill : The University of North Carolina Press,
[2023] | Includes bibliographical references and index.
Identifiers: LCCN 2023008284 | ISBN 9781469675985
(paperback) | ISBN 9781469675992 (ebook)
Subjects: LCSH: Audiocassettes—Social aspects. | Music—History and criticism.
Classification: LCC TK7881.6 .M37 2023 |
DDC 621.389/324—dc23/eng/20230302
LC record available at https://lccn.loc.gov/2023008284

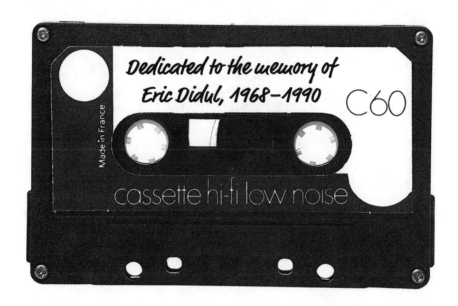

CONTENTS

INTRODUCTION

CHAPTER ONE

KILLING MUSIC

The Rise of the Cassette Tape

CHAPTER TWO

CREATING MUSIC

How Cassettes Helped Launch Movements

CHAPTER THREE

CASSETTES UNDERGROUND

An International Network of Tape Artists

HIGH BIAS

Introduction

Making, listening to, and caring for cassettes is *the most* hands-on and personal music listening experience. For sure. You don't just listen; you're very involved.
—**Adam Horovitz**, Beastie Boys

There are two things I remember well about my initial week of college. First, right after our inaugural dorm meeting in which we were warned about the dangers of alcohol, one of my dormmates started handing out cans of cheap beer. Second, a guy across the hall from me who would become my friend, Glen Springer, asked if I wanted a copy of a mixtape someone back home had given him called *Toxic Tunes*. I found the latter offer far more exciting.

Toxic Tunes was filled with songs by weird punk bands I had read about in high school but never actually heard. Where I grew up, it took an hour to drive to the closest record store, and that store certainly wouldn't have sold anything that was on this mixtape. Just the group names and song titles by themselves sounded illicit. Dead Kennedys, the

Meatmen, Butthole Surfers. "Too Drunk to Fuck," "Tooling for Anus," "Bar-B-Q Pope."

But then, cassettes had always felt a bit forbidden. When I started listening to music as a kid, buying a vinyl record album was approved, official, what you did if you wanted to hear something. Then I found out about tapes. Could I really copy albums from friends instead? Could I really put different songs from different records onto the same tape? Could I really dub one tape onto another? It all felt so down and dirty, so private yet so cool to share. And once I got to college, sometimes it seemed like making tapes was all I did.

> Wouldn't you really rather have a cassette than a record anyway? Cassettes don't scratch, they fit in your pocket, they're marvelously portable (home, car, friend's home, friend's car), and they stack up nice and neat. They also cost less.
> —**R. Stevie Moore**, from his R. Stevie Moore Cassette Club catalog

The cassette tape is revolutionary. It's small, it's cheap, it's easy to use. It's not necessarily more foolproof than a record—in fact you can screw it up even worse, and it can even just screw up on its own. But when a record gets scratched, it sounds annoying. When a tape warbles or flutters or wrinkles, it sounds . . . kind of cool? It makes you think, what if other music sounded that way? What if *my* music sounded that way? And you can fix a lot of tape problems yourself without knowing much about what you're doing. It has screws you can unscrew, spools you can wind with a pencil. You can even fix breaks in the tape—with tape! And you can always buy more blanks if the ones you have fail.

Tapes can go wherever you go. They can get lost at the bottom of your backpack. Huge sprawling piles of them can gather on the floor of your car. They can become orphaned from their cases and replaced by something that then, for some reason, never makes it back to its original home. They can shine from racks on your wall, their thick spines beaming the colors of handwritten titles toward you like light beckoning a moth.

The cassette tape is an audio medium that everyone can access and control and modify and remake and destroy and resurrect. It's an audio medium that was actually made for everyone. That's pretty revolutionary.

> Cassettes are one-to-one. That's the populist way. "Here's
> a tape." And you could just make the tape. So it was the
> people's format.
>   —Ian MacKaye, Fugazi/Minor Threat/Dischord Records

The cassette tape is a way people can talk to one another. Its uses and benefits and anomalies form a language—one that often brings another language with it, music. A mixtape you make for someone can be a code, a message, a signal, a conversation. Music you create and record on cassette can be a missive, a statement, a movement, a plea for attention. What you capture on cassette—concerts, songs from the radio, random noises—can become your hobby, your personality, your reputation. Sounds realign magnetic particles on a tape, and when you associate one sound with another because you put them together on a tape you listen to over and over, the tape realigns your brain.

Cassette tapes are personal, amateur, and subjective. They don't exist if someone doesn't hear them, and everyone hears them differently. They are for individual use and collective exchange. They have built communities, connected like-minded people over long distances, and passed along local and regional styles and innovations when no other means or medium would or could. They are do-it-yourself and do-it-ourselves. One person does something by themself on tape, and soon enough a bunch of people are doing it by themselves, together.

> It's sloppy, it's dirty, it's marked the way the human body is
> marked, by the space and time it passes through. It wears
> those scars and those scuffs, and that becomes part of why
> you love the tape.
>   —Rob Sheffield, from Cassette: A Documentary Mixtape

The cassette tape is imperfect. It degrades, it tangles, it adds noise, it adds hiss. It puts a smudgy fingerprint on everything it touches, and everything that touches it does the same. It eventually dies, though it often lasts longer than you expect.

For anyone who loves cassette tapes, its mechanics are magical. The way the case swings on a hinge like a miniature book, so satisfying to open and close. The way the cover, or J-card, folds into halves and thirds, with layers begging to be opened and perused. The way the outer shell protecting the tape is so smooth, molded, symmetrical. The way you can peek into the tiny window and watch the tape work, spooling forward and backward, or just let it sit still, waiting to be played, holding sound between its layers. The way the cassette tape fits about as perfectly as any object could in the palm of your hand.

> Tape has its own narrative, its own way of structuring narrative . . . and this is a narration intimately caught up with human belief in life as an accumulative narrative.
> —**Paul Hegarty**, "The Hallucinatory Life of Tape"

Cassette tapes are analog. They don't replicate sound exactly as it is. They distort it, and the more you copy them, the more distorted the sounds become. The story of the cassette tape is distorted too. You can sketch out a map of its journey, but the textures, the hills and valleys, depend on which trails you follow. Perhaps that's true of anything, but it's especially true of a format so customizable, so intimate and social and surreptitious. The cassette tape has meant so much to so many that its history is as diverse as the innumerable people whose lives it has altered.

What follows is a version of that history, tracing both how the cassette tape emerged—as a technological development, a marketed product, a cultural icon—and how things changed because of the cassette tape. It's a winding, messy path through international commerce, far-flung musical movements, covert underground cultures, and most important, intimate connections between people obsessed with their own ways of using and sharing cassette tapes.

In the technical lingo of cassette tapes, "high bias" means high quality. The higher the bias, the better the sound. The story of the cassette tape has bias, too. Every person who encounters a tape adds something to that story, whether by listening to it, recording over it, or passing it on. That's why this story is still going—because every cassette tape offers a chance to do something new.

> If a record sucks, it sucks. If a tape sucks, you can put something better on it.
> —**Mike Haley**, *Tabs Out* cassette podcast

CHAPTER ONE

Killing Music

The Rise of the Cassette Tape

The cassette tape has always been dangerous. Ever since it emerged in the early 1960s, it has been used to create, to invent, to individualize. In ways unlike any other audio format, the cassette tape offered freedom to artists, musicians, and fans—the kind of freedom that scared anyone used to dictating how music is made, sold, and heard.

Maybe you've heard the phrase "Home taping is killing music." Sounds pretty scary—but to the British Phonographic Industry, not scary enough. To get cassette tape users truly spooked, in the early 1980s the BPI created an advertising campaign that plastered this sensationalistic motto in bold block letters atop an ominous graphic. A cassette-shaped skull with its two holes serving as watchful eyes sat ready to pounce if you dared to tape music at home. Beneath were crossbones and another warning: AND IT'S ILLEGAL.

The British Phonographic Industry's
warning against home taping, printed
on the inner sleeve of an album by Kiss.
(Photo by Mark Lore)

The horror suggested by this creepy cartoon must have been confusing. Could the cassette tape really murder an entire art form? As
Newsweek put it in 1969, "One wouldn't think that the giant record
industry had anything to fear from a revolutionary only four inches
high." But now that people could copy music for just the price of a tape,
record companies were definitely frightened—not only in Britain but
around the globe. The industry tried everything—regulations, taxes,
court battles, public shaming—to quash home taping, which it saw as
taking money directly out of its pockets.

To those demonizing home taping, it wasn't just stealing. It represented a shift in the way music was controlled, an upending of the
hierarchy between producers and consumers. "Tapes . . . offer one dis-
tinct advantage that the record industry can't counter," wrote Richard
Harrington in the *Washington Post*. "They are reusable, adaptable to
the transient nature of music." In Harrington's piece, Lou Dennis of
Warner Brothers Records added, "As long as all the tape machines in

this country have a record button, how do you control that? You can't." Presaging the fears digital music would cause decades later, Irwin Tarr of RCA Records told *Newsweek*, "Young people now can start their music collections with tape and never have to buy a single record." "What is most frightening is that very soon it becomes a hobby," added Joe Cohen of the National Association of Recording Merchandisers. "And after it becomes a hobby, it becomes a habit."

"What the anti-piracy activists couldn't foresee was the intoxicating effect of assembling a mixtape or hearing yourself through your headphones," wrote essayist Hua Hsu decades later. "What the cassette introduced wasn't merely the impulse to copy and steal or to curate and create. The cassette inaugurated an era when it was possible to control one's private soundscape."

People could copy records and swap them with friends rather than having to buy everything they wanted to hear. They could tape albums off the radio, which sometimes played them in full without commercial interruption. They could compile favorite songs onto tapes, bypassing the way record companies delivered music and the way radio doled out hits. They could record their own creative work onto cassettes and release it on hand-dubbed tapes, eschewing conventional channels open only to a well-financed few. All these possibilities turned a simple physical object into the stuff of dreams. "[A blank tape is] a scramble of plastic, film, oxides, hubs, spindles. It's useless in itself," wrote Recording Industry Association of America president Stanley M. Gortikov. "It becomes valuable to its maker and its purchaser only when it comes alive and records our copyrighted music."

The home-taping panic began in the mid-1960s, just a few years after the cassette tape was invented. At that point, blank tapes were selling around 500,000 units a year in the UK alone, but by 1977 that figure ballooned to almost 40 million. That same year, 90 million blank tapes were sold in Germany, and by the early 1980s over 200 million were being purchased annually in the United States. This led to numerous industry studies, all intended to show the damage home taping caused. In 1977, the BPI claimed that 8 million people in the UK were copying over 80 million albums a year onto cassettes. Soon after, the BPI asserted that £100 million worth of sales was lost every year to home taping; in 1981, it said that figure had exploded to £305 million. In

America, one early 1980s study claimed home taping bled $1.5 billion a year from the industry; another cited a number two times that.

The rhetoric around these studies was often sensationalistic. Headlines spurred by industry organizations called home taping a "monster" and a "cancer." In a *Billboard* essay titled "Home Taping: Copyright Killer," Gortikov claimed to have met a brewery executive on a plane who said he never buys albums but instead has a "tape club" in which friends "copy whatever records we want." "For about every album we sold, one was taped," Gortikov added. "In our henhouse, the poachers now almost out-number the chickens."

The industry's arguments were framed not just as empirical truth but also as common sense. Why would someone buy a blank tape, they contended, if not to copy music? As journalist Adam White wrote in *Billboard*, "[Few] believe that such growth [in blank cassette sales] is attributable to more tape-letters being sent to Australia, or an upsurge in recording baby's first words." But some counterarguments made sense too: home tapers might buy vinyl copies of music they end up liking; they might tape music they already own for portable use; they might use tapes for all kinds of nonmusical purposes, including activities more common than "tape-letters," such as dictation and field recording. (There was also the inconvenient fact that prerecorded tapes were often more cheaply made than blanks, which sounded better and lasted longer.)

Some dealt with the problem by accepting it. A handful of stores rented records, though they often framed this as a way to try an album before buying it, rather than an encouragement of home taping. Island Records started a program called One-Plus-One in which they sold cassettes with one side of prerecorded music and another side blank, so consumers could use the second half for taping. "The public wants to home tape," said Island vice president Herb Corsack. "We can't fight it."

Most of the industry, though, reacted with panic. This increased with the introduction of dual-cassette decks, which could copy from one tape to another in a single unit. UK electronics company Amstrad helped turn this previously audiophile-driven technology into a cheaper consumer option. Its ads included a plea to buyers not to reproduce copyrighted material, but Amstrad founder Alan Sugar later admitted that he was using reverse psychology. "People would read it and think to themselves, 'Hey, that's a good idea! I can use this machine to copy my mate's ABBA cassette,'" Sugar wrote in his autobiography. By the mid-1980s, Amstrad blatantly touted its decks' dubbing capabilities. The BPI complained to the Advertising Standards Authority, which countered that it wasn't illegal to highlight the potential uses of "lawfully-constructed appliances." The BPI even took Amstrad to court, but a judge ruled that knowing equipment could be used to infringe on copyright doesn't make manufacturing the equipment illegal.

Industry associations attempted legal action as early as 1974 in the UK and continued to do so for decades. (Of course, these entities were just one part of the music industry; as Andrew Bottomley argues in his study of home taping, it's hard to say how much they represented actual musicians, if at all.) In the process, many solutions were proposed. One, an "amateur recording license" for which people could register and pay a fee, was unpopular in the industry because it seemed to legitimize home taping. It wasn't big with tapers either: Britain's 1976 attempt at such a program attracted only 5,075 applicants. Another suggested solution was inserting an electronic signal on albums that would be inaudible during playback but become "an unpleasant and irritating noise" when heard on a tape copy. Warner Communications even considered offering a financial reward to inventors of this imagined technology. Yet without specific legislation, manufacturers could still make recorders able to bypass any such signals.

One common proposal was taxation, usually as a levy on blank tapes, tape recorders, or both. US efforts in this direction got a boon from a 1981 court ruling that made taping television shows illegal. "The effect was to make instant criminals of millions of audio and video fans using their cassette machines to tape broadcasts," wrote Hans Fantel in the *New York Times*. "As it stands now, the situation is something like Prohibition in the '20s, when taking a drink made you an outlaw." The decision was overturned by the Supreme Court, which ruled that home taping fell under the 1976 copyright law's fair use guidelines. In 1982, a bill to tax tapes emerged in Congress, but it failed to become law, as did subsequent similar efforts. Other governments fared better: West Germany implemented a tape levy in 1979, and within a few years eight European countries as well as Australia passed similar taxes.

Those taxes generated small revenues compared to the losses the industry claimed, akin to putting a Band-Aid on a broken bone. Additionally, the question of how revenue was distributed, along with the cost of resulting bureaucracy, made taxes weak at best. For example, a BPI proposal would have increased the cost of a blank tape by 100 percent—generating £70 million to £80 million in revenue per year—yet it wouldn't be treated like a mechanical royalty in which much of the money goes to the artist. Instead, the proposal earmarked 40 percent of tax revenues for the record companies themselves.

Perhaps the most popular answer among industry associations was to increase public awareness—or, really, to browbeat consumers. This could have been framed as an appeal on behalf of artists, and at times the industry did trot out musicians to help. In 1981, the BPI's campaign began with the slogan "Home Taping Is Wiping Out Music," featuring recognizable artists such as Elton John and Gary Numan. A few years later, a set of American jazz musicians signed a letter inserted into vinyl albums, pleading that home taping threatened their livelihoods. In 1984, singer Beverly Sills wrote in a *New York Times* op-ed, "Why should song-writers, singers, musicians and artists who have devoted their lives to music, and enriched all of our lives as a result, be penalized and deprived of their right to be paid for their work?"

Ultimately, though, the industry's approach tended less toward encouraging sympathy for artists and more toward threats. Only weeks after the BPI campaign began, its slogan changed from "Home Taping Is Wiping Out Music" to "Home Taping Is Killing Music," launched in conjunction with hit-compiling record company K-tel. Not only did K-tel put the slogan on its next release, *Chart Hits 81*, but the company even directed sales staff to wear T-shirts bearing the warning and slap bumper stickers with it on their cars. A group called the Coalition to Save America's Music, which the Recording Industry Association of America's Goritkov claimed was "born out of fear—fear that home taping is bulldozing our copyrights, our jobs, our careers, our creativity," ran ads proclaiming that "Home Taping to Us . . . Is Like Shoplifting to You" and demanding people contact their representatives: "Write Them This Week. Or Else." BPI rhetoric was just as histrionic: "The record industry is evidently . . . a kind of cultural soup kitchen in which everyone may eat irrespective of their needs," said BPI director general John Deacon in 1987.

None of this resulted in much substantial change. Australia's tape levy and a proposed UK tax were both ruled unconstitutional; a 1985 US home-taping bill that included taxes died among lawmakers' skepticism about how the money would be doled out. And despite the scare tactics, home taping was never actually deemed illegal. In America, it was protected under "doctrine of first sale," which meant that once you bought a record, it was yours to do what you wanted with. In the UK, a big obstacle was enforcement: How could you catch someone taping an album other than by entering their homes? BPI press officer Richard Hobson rather sinisterly told the *New Musical Express* that "there are ways of getting access to people's premises provided one has reasonable grounds for suspecting they are breaking the law." But legislators and the industry ultimately decided that such privacy infringement was worse than theoretical lost album sales. "You'd have to be banging on doors and arresting people," said Warner Brothers' Lou Dennis. "And you can't do that."

Still, in some ways, the home-taping panic worked. A perception that it was illegal or at least immoral took some hold, even if few were punished for it. But it also succeeded in making tapes seem even cooler and more rebellious than before the industry targeted them. Over ensuing decades, the "Killing Music" slogan became a frequent punch line for independent-minded artists. In 1981—mere weeks after the BPI debuted the motto—the San Francisco punk label Alternative Tentacles released a cassette of Dead Kennedys' *In God We Trust, Inc.* with a message on the shell: "Home taping is killing record industry profits! We left this side blank so you can help." The metal group Venom wrote "Home Taping Is Killing Music; So Are Venom" on their 1982 release *Black Metal*. Peter Principle offered "Home Taping Is Making Music" on the back of his 1988 album *Tone Poems*. In 1994, Flying Saucer Attack etched a message on their LP *Further*: "Home Taping is Reinventing Music."

Most famously, in 1980 British band Bow Wow Wow released what was possibly the first cassette single. One side was blank to use for recording, but the other side was even more subversive, containing a song that touted home taping. "I don't buy records in your shop / Now I tape them all, 'cause I'm Top of the Pops," the band sang, later fantasizing about confronting cops who arrest them for home taping by "blowing them out with my bazooka." Bow Wow Wow's label, EMI, supposedly refused to promote the single—though it charted anyway—and dropped the group in part because of their message.

Somewhat lost in these arguments about home taping was the fact that it had been possible for decades, on a variety of formats, before cassette tapes made it controversial. "The audio tape recorder has been marketed for thirty years," wrote Jack Wyman, an executive for the Electronic Industries Association, in a 1982 issue of *Billboard*. "Yet the recording industry less than four months ago introduced for the first time legislation that attempts to transform home audio taping into an infringing activity." Why? Because the cassette represented something revolutionary in audio recording—even if its creation took a long, winding path.

The journey to the invention of the cassette tape began with the introduction of magnetic recording. Until then, recording sound on a phonograph was an acoustic process. The air pressure created by sound moved a needle that carved out an "analog" to that sound on a disc, which could then be played back by a needle. Eventually, the process became electric, as sound was captured by microphones and converted to a signal. But that signal was still mechanically etched into an analog on disc.

Magnetic recording offered an alternative to disc cutting. The idea goes back at least to 1878, when mechanical engineer Oberlin Smith visited the laboratory of Thomas Edison, who had recently secured a patent for the phonograph. Smith was intrigued by the potential in Edison's invention but also by one of its flaws: during recording, friction between the needle and the disc added noise. Looking for a way to avoid this problem, Smith imagined transferring the electric signal to a magnetic object by realigning its magnetism, a process that would not require physical contact. He wrote about this concept and worked on it but was never able to develop it much beyond dream stages.

Twenty years later, a telephone engineer in Denmark named Valdemar Poulsen had a different problem. He was frustrated that he couldn't leave a message when he telephoned someone and got no answer. He knew about Smith's magnetic recording concept and pondered using it to create some kind of answering machine. As a trial, he pulled the microphone out of a telephone, talked into it, and fed the resulting current to an electromagnet, which he moved along a piece of piano wire. When he connected that electromagnet to the earpiece of a telephone and moved it along the piano wire again—converting the magnetic pattern he had created back into an electrical signal—he heard his own message. He refined this process to create a recorder he called the Telegraphone.

Poulsen's method came to be known as wire recording. It took a while to catch on, but home recording already had its proponents using a variety of mechanisms. In the late 1930s, acetate discs offered a way to record audio on Edison's phonograph, but they could only be played a few times before wearing out. In the early 1940s, dictation machines that recorded onto disc were somewhat common, as was the Recor-dograph, which used a needle to record audio onto 35-millimeter film and was sometimes carried by reporters during World War II. But all of these machines were difficult to use and easy to mess up—recurring problems with almost every recording technology that preceded the cassette tape—and they all faded away.

Eventually, wire recording gained some steam via machines that used stainless steel wire the size of sewing thread. By the mid-1940s, these recorders sold in the tens of thousands in America. But there was a catch, literally: the wires would often twist, tangle, and snag, frustrating users and sometimes destroying recordings. A new material was needed that could be magnetized and used repeatedly without breaking so eas-ily. Fortunately, an inventor named Fritz Pfleumer found a solution while solving a different problem. His innovation grew from his work in the tobacco industry. In the 1920s, German cigarette makers sought to adorn the ends of cigarettes with something cheaper than the gold-leaf band they currently used. Pfleumer suggested using powdered bronze by coating it onto the existing cigarette paper rather than adding a separate band. Pfleumer already knew about Poulsen's recording method and figured if he used his coating process to add magnetic material (in this case, iron oxide) to paper, it might work better than wires. He patented this concept, calling it "sounding paper," in 1928.

Over the next two decades, Pfleumer's magnetic-coating idea was improved and expanded upon, overtaking wires as the preeminent audio-recording medium. The format evolved from paper to plastic tape, which produced less noise. Working with Pfleumer, the German elec-tronics company AEG created the Magnetophone, the first prominent tape player and recorder. Its primary mechanical elements were two rotating hubs and a play/record head. A reel of tape would be placed on one hub, then threaded so it passed over the head. The tape was then wound onto an empty second reel, so this kind of machine became referred to as "reel-to-reel." Similar magnetic recorders were created around the world near the same time, including Britain's Blattnerphone and America's Soundmirror. But the Magnetophone was cheapest, most portable, and most reliable and came with its own built-in speaker. Early versions didn't sound great, but improvements came soon, both in the tape materials (with gamma ferric oxide becoming the standard) and the recording process.

One adjustment consisted of adding an inaudible high-frequency tone during recording, a process engineers call "biasing"—in this case alternating current, or AC, bias. (Poulsen had used a version of this, direct current, or DC, bias, when working on his wire-based Telegraphone.) By reducing distortion and noise, biasing so improved the quality of recordings that eventually blank tapes would be touted for their ability to handle levels of bias—the greater the bias, the better the sound—with "high bias" tapes being the best you could buy.

The Magnetophone's circulation was small at first. Only about 200 were used in Germany in 1937, three years after it was introduced, primarily for office dictation. Eventually it was adopted by the military, the government, and broadcasters. Once it was marketed to consumers in the 1950s and introduced in America, it grew so popular that in several parts of the world the name Magnetophone became synonymous with "tape recorder." Still, reel-to-reel tapes struggled to become a popular consumer product rather than an audiophile toy. Prerecorded tapes, of which there were few, could be three times as expensive as records. And playing a reel-to-reel tape was difficult and time consuming. It took some training to learn to thread a tape onto a player. "As we use equipment, we more or less are all fools," explained Dutch engineer Lou Ottens in Zack Taylor's 2016 film *Cassette: A Documentary Mixtape*. "The reel-to-reel was absolutely not foolproof because you could mishandle it in all sorts of ways. You can't go on with products that ask for specialists to handle them; everybody should be able to handle it."

The key to making tape easier to manage was enclosing it in some kind of case, so that it would thread itself and the user wouldn't have to touch it. This idea predates magnetic tape. In the early 1930s, a German company marketed metal shells that contained recording wire. Soon known as cassettes (a French word meaning "small cases"), they never caught on beyond specialist uses. In the 1940s and '50s, some tape-based cassette systems emerged, but there was a big problem. To provide decent-quality audio, the tape had to travel at high speeds; low speeds introduced problems of frequency response and pitch fluctuation known as wow and flutter. High-speed recording used a lot of tape, so cassettes had to be big and bulky, making portability difficult.

The first successful enclosed recording product arrived in the 1950s, in the form of a cassette with a single reel. Tape fed from the inside of the reel to the play head and then back to the outside of the reel, so the audio could start over without being rewound. This endless-loop system was devised by inventor Bernard Cousino in 1952, and cassettes that used it became known as cartridges. Two years later, Cousino's coworker George Eash adopted the idea to create the monaural Fidelipak. It became the first widely used commercial cartridge, primarily adopted by radio stations, since the endless loop was perfect for oft-replayed advertisements.

In the early 1960s, California car dealer Earl "Madman" Muntz heard about Fidelipak and worked toward a version that could be played in automobiles. He touted his Muntz Stereo-Pak as high fidelity (or "hi-fi") because its four tracks could play stereo audio in pairs. He also made it forty minutes long to encompass an album's worth of music. William Lear, owner of Learjet, began using Muntz's cartridges in his planes. But he wanted longer tapes, so he had engineers create a cartridge that could hold eighty minutes of audio and included four sets of paired stereo tracks (twice as many as the Stereo-Pak). This became known as 8-track tape. Its ease of use and small size helped it catch on, especially after Ford Motor Company installed Lear's player in its cars and record label RCA Victor released albums on prerecorded 8-tracks. Over 2 million 8-track players were in use by the late 1960s.

But 8-tracks and other cartridges still had drawbacks. They couldn't be rewound by the user, and forwarding was limited, especially compared to the way a listener could jump to any point on a vinyl album. Reel-to-reel tapes didn't have this problem, so the solution could be a product that enclosed tape but contained two reels rather than one. RCA first introduced such a cassette in 1958, with the added innovation that the hubs that wound reels didn't have outer rims to shield the tape, what were known on reel-to-reel machines as flanges. The cassette case already provided such protection. Flangeless hubs could sit closer together, making for a smaller cassette.

The RCA cassette was still rather large though—five inches tall by nine inches long—because it had to include enough tape for decent-quality audio. Additionally, the power consumed by RCA's player meant it couldn't run on batteries for long, so it wasn't very portable. Around the same time, CBS created a cartridge that used thinner tape and slower speed for use on improved play/record heads, meaning it needed less tape and thus was smaller. But it was a one-reel system, too unwieldy and expensive for nonprofessional users.

This is where the key figure in the history of the cassette tape came in. Starting in the 1950s, engineer Lou Ottens worked at Philips, a company formed in the Netherlands in the late 1800s. At the Philips factory in Hasselt, Belgium, Ottens led a team of developers aimed at the consumer recording market, and he dreamed of a portable, easy-to-use tape player. He even carved a piece of wood of the size he thought would work best and carried it around in his pocket. In the early 1960s, "we made a working sample of a tape deck based on a shrunken sort of RCA cartridge with twenty minutes playing time and the CBS size of tape," he told the *Register* in 2013. "It worked surprisingly well." His group was buoyed by the fact that in the late 1950s, Philips had sold over a million of its own battery-powered reel-to-reel machines. "It made us confident that there would be a big market for a smaller, pocketable battery tape recorder," Ottens said.

Ottens would later recall his primary design goals at a 1966 Audio Engineering Society convention: small dimensions, simple construction, reliability, durability, and low energy consumption. One thing he didn't mention was audio quality. His team assumed its cassette would be used to record simple sound that didn't require high fidelity. "It was initially presented as an opportunity for journalists, or nature lovers to make sound recordings outside," says Willy Leenders, a member of Ottens's crew. "The very first one, we said, 'Well, speech is good enough,'" Ottens recalled in *Cassette: A Documentary Mixtape*. "Then we came to the conclusion that [the sound quality] was much better than we had anticipated."

In 1963, Philips introduced Ottens's "compact cassette" and accompanying "pocket recorder." In its first year, this invention didn't make a huge impact, selling only 9,000 units. "The device was not at all welcomed as a revolutionary breakthrough," says Leenders. "It was initially not easily available either." But it didn't take long to stir up international interest. After the pocket recorder was introduced at the Berlin Radio Exhibition in late 1963, imitations popped up around the world, especially in Japan. In Leenders's book about Philips, *A Hystory of the Future*, he recalls that one of their recorders went missing at the Berlin expo, and later a Japanese version appeared that was so similar it even replicated a small design flaw in the Philips prototype. "We went to Japan and said, 'Gentlemen, if you want to imitate us, we better standardize, because without a standard, it becomes a mess in the whole world,'" said Ottens in *Cassette: A Documentary Mixtape*.

According to Japanese company Sony, Philips initially offered to license the compact cassette design for a fee, and when Sony threatened

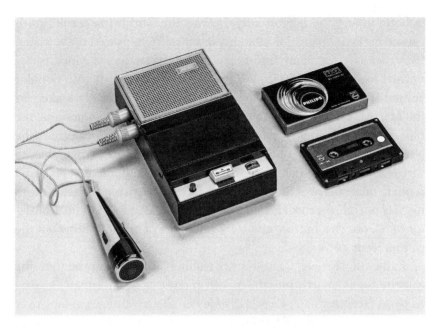

The first Philips compact cassette recorder, introduced in 1963. (Photo by Royal Philips)

to use another company's design instead, Philips relented and waived royalties. The Dutch company decided to do the same for manufacturers around the globe, granting its license for free to anyone who promised to follow its quality-control guidelines. Philips hoped that standardization would increase spread of the format and produce more sales than if the company kept the technology for itself.

This decision, perhaps even more than Ottens's original invention, was the big bang for the compact cassette. It would quickly become the dominant tape format worldwide, and soon when anyone said "cassette tape," they meant the compact cassette. Within five years of its 1963 debut, boosted by the success of Philips's Carry-Corder (marketed under its Norelco brand) in America and Sony's Magazine-Matic model in Japan, nearly 2.5 million cassette players had been sold around the globe by eighty-five different manufacturers. In 1967, the Carry-Corder cost $69.95, and a blank tape long enough to hold two albums cost $3.75; just a year later, tape recorders retailed for under $30.00.

Record companies made some albums available on prerecorded cassette as early as 1965, and by the end of the decade they were all in on cassette tape. "The biggest surprise was the worldwide revolution it caused in the individual availability of music," Ottens told the *Register*. "But that surprise came into being only very gradually, which is not normal for a surprise." The format made quick inroads into popular culture too. In 1968, a *Business Week* article titled "Music Maker for the Masses" declared, "That expensive plaything of the audiophile, the tape recorder, is finally entering the mass market," adding that "the teenager—the major market for recorded music—no longer has to thread a tape through a bulky and costly piece of equipment in his living room to make his own music. Instead, he can snap a blank cassette into his tiny portable recorder, stand its microphone in front of his transistor radio, punch a button, and copy two hours of music, editing out the commercials as he goes."

In the 1969 article "Cassettes Are Rolling," *Newsweek* raved that "the cassette snaps smartly into the mouth of any cassette machine like perfectly fit false teeth. . . . Cassettes enjoy all the advantages of tape over records—they are durable, compact, do not collect dust, and last almost indefinitely." Even *High Fidelity*, a magazine dedicated to the audio quality reflected in its name, was excited about this new, non-hi-fi format. "At last, here is a tape machine that is easy to operate, that can play under a wide variety of listening conditions and locations, for which there is an obviously growing pre-recorded repertoire of all types of music—and which you can also use as an all-purpose recorder," Robert Angus and Norman Eisenberg wrote in the magazine in 1969. "The combinations possible with a cassette device seem limited only by the imagination of the equipment manufacturers—and rarely in the history of home entertainment equipment has this imagination been so abundantly evident in such a short time."

In all these raves, sound quality was rarely discussed. Though Ottens and Philips had seen the compact cassette's potential as a professional format, its audio was generally considered rather low in fidelity. Reasons included wow and flutter, limited frequency response, distortion of the original audio signal, and the most common complaint, tape hiss. Caused by particles that are too thick to become fully magnetized during recording, tape hiss is a sibilant sound that floats over and sometimes even masks the original audio. Improvements came relatively quickly on many fronts. New tape materials and magnetic coating (chromium dioxide at first, then more expensive metallics, and finally ferric oxide mixed with cobalt) helped, as did refining the design of play/record heads. Even better was noise reduction from encoding techniques such as the Dolby system, which claimed to reduce tape hiss by 90 percent and eventually became common on most prerecorded tapes and tape players.

Still, few who latched on to compact cassettes were concerned with state-of-the-art sound. There aren't many stories of people rushing out to buy the newest metal-based tape or even much indication that most consumers knew what "bias" meant. Some actually embraced the technical shortcomings: the way sounds get murkier and more distorted as a tape is copied over and over, the way tapes degrade over time, the way you can hear echoes of the music you record over on a used tape, a phenomenon known as bleed through. "The cassette sound—whether heard at home on a stereo deck, in a car, or on a beach with a blaster—lends itself to so many levels of lo-fi or hi-fi that it is possible to do anything on them," wrote musician Eugene Chadbourne in Robin James's tape-essay compendium *Cassette Mythos*. "The most common point of view is: 'Who gives a shit? If you don't like it, dub over it.' After more than a decade documenting my music, that's the most exciting thing I've heard."

Instead of advances in audio quality, what pushed the cassette tape further were improvements in convenience, portability, speed, and all kinds of things that make it easier to listen to music, make music, and spread music around. Take the introduction of "tabs," two small plastic squares at each end of the top of a cassette. By breaking these tabs off, you create a hole that a deck can detect, causing it to disable its recording function and prevent erasure. Conversely, on prerecorded tapes, there were only holes—but if you put a little adhesive tape over them to trick a deck into thinking the tabs were back, you instantly had a new "blank" tape.

Novel features were also continually added to tape players and recorders that gave users new ways to take advantage of cassettes. The dual-cassette deck allowed people to directly copy one tape to another; eventually this could even be done at high speeds, so it would take less time to dub a tape than to listen to it. This was not only a boon to people who wanted to share albums but also a huge opportunity for self-recorded artists to make copies of their work to sell, allowing them to essentially become their own record labels. Auto-reverse tape decks allowed listeners to avoid taking out a tape and flipping it over. When one side ended, the play head switched direction, so the next side played automatically. Multicassette carousels and changers encouraged users

Philips engineer Lou Ottens holds his invention, the compact cassette, in 1988. (Copyright Royal Philips / Philips Company Archives)

to load in stacks of tapes and listen to them all in a row. Gap detection let listeners fast-forward a tape automatically to the next song, with the deck recognizing silence between segments of audio. Philips even made a deck that allowed users to "program" an existing cassette by adding small, inaudible signals to chosen locations on the tape. When you punched in a set of numbers corresponding to those signals, the deck would forward and rewind to songs in the order you chose, a precursor to user-created playlists in the digital streaming era.

Helped by all these changes and improvements, the popularity of cassette tapes continued to rise. By the mid-1970s, Philips had produced a million cassette recorders and 50 million cassettes. In the early 1970s in the United States, vinyl album sales topped cassette sales by almost twentyfold; by 1981, that factor was just around two. Only a year later, cassettes overtook LPs in sales, something the *New York Times* described as "the climax of a Cinderella story in which the lowly triumph against all odds." In the twenty-five years after Philips introduced the compact cassette, the company claimed to have sold over 3 billion tapes. "We expected it would be a success," said Ottens in *Cassette: A Documentary Mixtape*. "But not a revolution."

Some advancements in cassette tape technology transcended commercial popularity, significantly changing the way music is made and consumed. In 1966, Philips introduced the 22RL962, a player/recorder that contained a cassette deck, a radio, and a speaker. It could plug into a wall or run on batteries, meaning it was portable, though too big for pockets. The 22RL962 seems like a logical enough step in the development of cassette machines, but two specific features helped its success grow in ensuing decades. First was the size of the speaker, larger than any on previous players. Its power meant this wasn't just a portable tape player; it was almost like taking your sound system out of your house and into the world. Second, it could record audio directly from radio to cassette tape, a process that had often required multiple machines and connecting cords.

As with the compact cassette itself, the Philips 22RL962 intrigued audio companies in Japan, who seized on the concept and ran with it in the 1970s. This explosion was due in part to a wave of young Japanese citizens migrating from country areas into cities, where living spaces were small. They wanted high-fidelity audio systems that didn't take up much space but could fill any room with music. Manufacturers responded by expanding the Philips model, crafting larger devices with speakers that could blast out a wide range of sound, especially low-end bass tones, at high volumes.

Many of these portable sound systems made their way from Japan to the United States, and one model in particular caught fire. In the mid 1970s, the Victor Company of Japan, or JVC, introduced the RC-550. With large speakers, including a ten-inch woofer supplying bass sound, a huge handle, and even a shoulder strap, it was a massive device, both in terms of how loud it could sound and how big it looked. On the streets of New York, it became known as El Diablo. Within a few years, JVC introduced another model called the RC-M70, whose stereo speakers boasted forty watts of power, making it equal to or better than home stereo systems. It was huge and heavy (almost twenty pounds), in part because the casing needed to withstand all the vibration from the shaking bass. It's likely that the RC-M70 was the first of these machines to be referred to by a name that would stick for all of them: the boom box.

In this case, "boom" meant bass, and a musical style emerged in New York in the late 1970s that was perfect for a device that could deliver lots of low end. Hip-hop was built on heavy bass and big beats, and boom boxes became the way that sound got around. Young people brought their increasingly bigger, louder, and more ostentatious decks out onto corners and blasted music around blocks, turning the boom box into a street status symbol. Lyle Owerko, author of *The Boombox Project*, likened the device to "a sonic campfire, with people gathering around to generate dialogue, debate, heat."

"That changed music," Owerko told NPR in 2009. "[It] empowered them to say, 'I want to go into my bedroom and record something and then bring it back on the street so that you can hear what I want to say, and I'm going to say it to music, and I'm gonna play that music, and you're gonna like that music, and if you don't like that music, I'm going to play it louder, and I'm gonna make it so I don't disappear and what I want to say doesn't disappear.'" In the same piece, hip-hop historian Fred Brathwaite (a.k.a. Fab 5 Freddy) concurred: "A big part of this hip-hop culture in the beginning was putting things in your face, whether you liked it or not. That was graffiti, or a break dance battle right at your feet . . . or this music blasting loud, whether you wanted to hear it or not."

Some people definitely did not want to hear it, and the boom box quickly became a flash point for cultural debate. Media reactions ranged from bemusement (why would anyone want to lug something that big around or listen to music so loudly?) to outrage over the social disruption caused by broadcasting your choice of music to everyone around you. A 1980 *Wall Street Journal* op-ed even claimed that boom boxes were usually carried by men "eight feet tall and weigh[ing] four hundred pounds" who walked around with "a knife in one pocket and a zip gun in the other." Far from deterring the popularity of boom boxes, those who decried the shirking of unwritten rules in public spaces pumped up the defiant appeal of a device that already had a lot of rebellious cachet.

Just as important, the boom box's public quality created a social interchange in hip-hop, similar to the way folk songs were passed around verbally before the advent of recording. "Music went from home collective to public collective," pioneering hip-hop DJ Robert "Bobbito" Garcia told Owerko. "[It] took a step from being in somebody's apartment to cats literally taking their speakers and turning them outside their windows. These were people who were not DJs, who were just sharing music. . . . This was just 'I love music, and I'm sharing it with my peers.'"

Of course, any portable device with a large speaker could have filled this new, outwardly social role. After all, teenagers had been cruising the streets with their car radios blasting for years. What made the boom box stand out was its cassette deck. Now, people had a way to play music they controlled, even music they made, out in public. Hip-hop in particular thrived on cassette, primarily via tapes of parties where DJs mixed the latest beats and raps into aural collages. These tapes were dubbed, traded, sold, and heard, influencing a growing circle who could listen on boom boxes. "The boom box was the actual conduit to how we communicated the music," rap pioneer Kool Moe Dee told Owerko. "[It] would be the only way you would actually hear hip-hop."

In the early 1980s, New York was just starting to recover from a period of desolation and vacancy, and the boom box provided a new pulse shaking the streets. "While the urbanscape was crumbling, the audio soundtrack of the boom box cut through the crime and the grime, calling out for change," wrote Owerko. "Recognition of that call's urgency echoed around the world." And as the boom box got shinier and flashier—sporting blinking colored lights and all kinds of dazzling functions—it resembled, as film director Earl Sebastian told Owerko, "a New York skyline. . . . The boom box spoke New York. . . . Its sound was having your own piece of New York." Hip-hop and the boom box became nearly synonymous. When LL Cool J put a JVC RC-M90 on the cover of his 1985 debut album *Radio*, that instantly told you what kind of music you were about to hear.

The boom box's influence wasn't confined to hip-hop or New York. Numerous other genres and subcultures—particularly underground rock and heavy metal—seized on the device, especially as a tool for recording. Bands stuck their boom boxes in the middle of practice spaces (garages, basements, bedrooms) and banged out their music onto tape, sometimes releasing the results as finished product. In the process, boom boxes became instruments themselves. "I had one boom box that had perfect distortion," says Alan Bishop of experimental rock group Sun City Girls. "I tried to replace it with the same model, but for some reason the input wasn't messed up just enough to where I could get that distortion. A lot of my friends had a favorite boom box like that, that would add something that just came out perfectly."

Ultimately, the appeal of the boom box was more universal than any single genre, scene, or culture. Like the cassette tape itself, it offered a chance at independence and personality—a chance to listen to and make your own music, not someone else's. "My boom box really was sort of an initiation into manhood," writer Andre Torres told Owerko. "It really gave me my independence and the ability to have my music with me when I wanted it."

One particularly keen observer of the boom box phenomenon was Akio Morita, cofounder of Sony. "In New York, even in Tokyo, I had seen people with big tape players and radios perched over their shoulders blaring out music," Morita wrote in his autobiography. "I knew from my own experience at home that young people cannot seem to live without music." This led Morita and Sony to create something that would soon challenge the boom box's prominence in the realm of portable cassette technology.

According to Morita, his Sony cofounder Masaru Ibuka came into his office one day in the late 1970s with a quandary. He wanted to listen to music on the go without subjecting other people to it or carrying around something as big as a boom box. Intrigued, Morita directed his engineers to modify an existing Sony product—the Pressman tape recorder, used mostly to capture speech—by replacing its recording capability with a stereo amplifier. He also requested that Sony come up with lightweight headphones more portable than their current models. Playing off the Pressman name, someone at Sony dubbed Morita's idea the Walkman. Though Morita didn't love the handle and tried others (including Walking Stereo, Stowaway, and Soundabout), the original moniker stuck.

Alternate stories of the Walkman's origin abound. In his history of Sony, author John Nathan reports that Ibuka sought a way to listen to music on long international flights and had the engineering department create a solution before Morita even heard about it. Some say Morita thought of the Walkman himself to use to tune out his noisy children; others claim the Sony tape division, nervous about being eliminated, rushed to come up with something new to justify its existence. (Morita admitted to encouraging multiple stories, happy to see a legend grow around the device.) A German businessman named Andreas Pavel swore he came up with the same idea and a primitive prototype before Sony did—and this story turned out to be true. Though Sony apparently had no knowledge of his work when it made the Walkman, Pavel eventually got a settlement from the company after many years of lawsuits.

All this mythology may sound odd for a device that seems so logical now. After all, the Walkman simply took something that already existed—a portable, pocket-sized tape player—and added something else that already existed, headphones. But at the time, it was received with both wonder and skepticism. In her fascinating book about the Walkman, *Personal Stereo*, Rebecca Tuhus-Dubrow quotes Pavel's stories about his pre-Walkman contraption blowing his friends' minds. "[It] put us in a state of ecstasy," he said. "We started feeling as if we were floating through the trees. It was unreal. . . . Life became a film. . . . It actually put magic into your life."

On Sony's end, Morita had to battle the doubts of his own employees, especially engineers who couldn't imagine why anyone would want a tape player that couldn't record. He kept fighting for the Walkman, though, touting this persistence in his autobiography. "Our plan is to lead the public with new products rather than ask them what kind of products they want," he wrote. "The public does not know what is possible, but we do." Yet the public had some say, too. Early on, Morita insisted the Walkman include two headphone jacks and a device called a hot line, allowing listeners to pause audio and talk to each other. His assumption that people would use the device communally rather than individually turned out to be incorrect. Almost immediately, Walkman users reveled in listening alone and being enveloped by music. As one early adopter told the *New York Times*, "It shuts out the awful sounds of the city."

In the summer of 1979, the Walkman arrived in Japanese stores. Sony promoted it in part by giving it free to some young people, instructing them to use it while strolling city streets. The first run of 30,000 sold out by September; about a year later, 2 million units had flown off shelves. In America, the Walkman's tech-forward novelty—and $200 price tag—made it as much status symbol as listening device. In a *Wall Street Journal* article with the derisive headline "New Cassette Player Outclasses Street People's Box," a Bloomingdale's manager called the Walkman "a unique product for personal gratification that appeals very much to the sophisticated." One *New Yorker* writer described spotting the device around Manhattan as if reporting on fashion trends. A man on the street in a pinstripe suit described it to the *New York Times* as "the thinking man's box," while another claimed users would nod knowingly on sidewalks "like Mercedes-Benz owners honking when they pass each other on the road."

But as it and other similar devices by Sony competitors became smaller and cheaper, the Walkman became less for "the thinking man" than for everyone. Its ubiquity on streets, on public transportation, and at gyms led it to garner nearly as much criticism—and symbolize as much rebellion—as the boom box. Though headphones were much quieter than boom box speakers, some still complained about the noise. In England, the London Transport implemented regulations for how loud a Walkman could be played, charging fines to violators. Others derided the supposed antisocial, even anti-intellectual nature of the Walkman. A 1981 *Money* article described Walkman users as "glassy-eyed folks with . . . little boxes hanging around their necks [like] members of some crazy new cult"; pompous reactionaries such as Allan Bloom declared that "as long as they have the Walkman on, they cannot hear what the great tradition has to say."

Rather than discouraging Walkman users, this kind of rhetoric emboldened them to view their personal listening habits as acts of independence. "I know how people who don't wear Walkmans feel about the rest of us," wrote an anonymous author in the *New Yorker* in 1989. "I know because they ask me if I think it's a good idea to wear headphones around, as if there might be something natural or wholesome about subjecting oneself to the cacophony of, say, a midtown sidewalk next to a construction site during rush hour. . . . But when I'm listening to the Walkman, I'm not just tuning out. I'm also tuning in a soundtrack for the scenery around me." In the British music magazine *Touch*, editor Vincent Jackson mocked the dirty looks he got when wearing his Walkman on public transportation. "It's not the sound coming out of your headphones that bothers people around you, it's the symbolism," he wrote. "When your average commuter spots a young person nodding his or her head to the beat of the Walkman, they immediately connect the personal stereo to all that represents the insubordination of youth."

Jackson wrote those words many years after the Walkman's debut, when the device was still going strong. A decade in, Sony had manufactured 50 million units; by 1995, that number had tripled. Over time, features were added such as a radio tuner, recording capability, waterproofing for swimming users, and more. As CDs eclipsed cassettes, the Walkman morphed into the portable CD-playing Discman. As digital

Sony cofounder Akio Morita with
various editions of the Walkman in 1989,
ten years after the device was introduced.
(Copyright Sony Group Corporation)

media overtook CDs, Apple's MP3-playing iPod became essentially the
new Walkman. Similar in size and shape, it was valued for the same
qualities: its portability and individuality, a conduit for making your
own soundtrack to your life.

Still, the Walkman's impact was primarily due to the cassette tape.
After all, portable radios with earpieces already existed, and few
treated those as life altering or society destroying. The Walkman was

revolutionary because it was personal, just like cassette tapes. You were no longer bound to what songs came on the radio or an album; whatever you wanted to hear could be put on one small object fixed to your clothes. The Walkman was like a new appendage, and it helped turn music listening from a pastime into an act of self-definition.

Around the time that the Walkman was born, another company in Japan introduced its own influential tape device. This one changed a different side of music: the way it gets made. The concept of "amateur" recording—capturing music, voice, or any other sounds by yourself—goes back to the early days of Edison's phonograph. Once magnetic recording took hold, home recording on reel-to-reel tape recorders spread. But because of the expense and expertise required, it was primarily the domain of professionals and serious hobbyists.

That shifted in 1979, when the Tokyo Electro-Acoustic Company (TEAC) introduced the Tascam M-144, also known as the Portastudio. Based on one of TEAC's reel-to-reel recorders, the Portastudio took advantage of the four tracks on a cassette (two stereo tracks on each side) to make multitrack home recording possible without using reel-to-reel tapes. Users could plug in a sound source like a microphone or a guitar, record onto one track, then rewind and record something else onto the next track while listening to the previous recording. To include more than four sound sources, they could use the device's built-in mixer to unite everything recorded so far onto one track (a process known as bouncing or mixing down), freeing up those previously used individual tracks for more recording.

The Portastudio was relatively small—just eighteen inches wide and fifteen inches long, weighing less than twenty pounds. It was also decently affordable, selling for around $1,100. But the potential it represented was huge. Before, if a band wanted to make a record, they had to pay for expensive time in a studio or record all their instruments at once with a boom box or other kind of tape recorder. But now they could do professional-style multitrack recording cheaply at home. Soon, many companies besides TEAC—most notably Fostex and Yamaha—offered a four-track cassette recorder, eventually referred to simply as the 4-Track. At the time, cassettes were still considered sonically inferior to the larger, faster-speed tapes used in recording studios, so manufacturers were hesitant to call the 4-Track "professional." In the *Billboard* announcement for the debut of TEAC's M-144, a sales manager gushed about the device's many uses but quickly added, "What the Portastudio is not, however, is an audio high fidelity product." A slogan in an early Yamaha ad, "Go to your room and play," made the device seem like a toy.

But excited adopters of the 4-Track cared little about the sound quality and a lot about how they could use the device. Some sketched out "demos" of music that would later get re-recorded in a studio. Others used the 4-Track to make final recordings, dubbing copies onto blank tapes to pass around, trade, and sell. Hip-hop DJs used them to make mixtapes of music they played in clubs, which they then sold to people who wanted to hear the music again—or who had missed the party and needed to hear what all the fuss was about. Underground artists made sound collages on 4-Tracks and exchanged them through the post office as international mail art. Rock bands recorded full albums for independent labels on 4-Tracks and sold reams of cassettes without ever having to set foot in a studio. "Bedroom" musicians recorded songs in the middle of the night using headphones, suddenly able to make music without disturbing, or dealing with, the world around them.

In all cases, the 4-Track offered freedom in both the creation and dissemination of music. "I record on a cassette and copy on cassettes—the whole means of production stays in my hands," musician Hal McGee told the *New York Times* in 1987. "And I have reason to believe there are thousands and thousands of other people doing it, too. They're not waiting around for the big recording companies to tap them on the shoulder and give them the right to communicate with the rest of the world. Doing this, you're not going to get 10 million people to hear you—but you can do what you want." This appeal survived well into the digital era. "The 4-Track was like an icon," says Brian Weitz of Animal Collective, a band formed after the turn of the century. "It was liberating to a teenager, to feel like the gates aren't as impenetrable as you think they are."

As four-track recording spread, the allegedly inferior audio quality of cassette tapes became a badge of honor. The inconsistencies, distortions, and hiss of tapes were fitting for artists whose intent was not to make money or sell products but to communicate directly. Using imperfect technology didn't automatically grant an artist authenticity, but those willing to forgo fixing what some would see as "flaws" in their music were surely less concerned with commercial viability than artistic expression.

"Doing things wrong, you kind of turn them into your own version of doing things right," says James McNew, who started recording his own music on a Tascam 4-Track in the late 1980s. "I loved the fact that if I

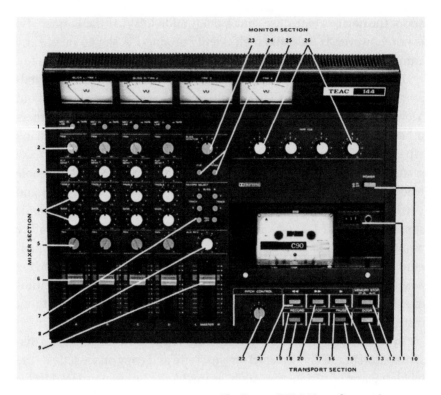

The Tascam TEAC M-144 four-track recorder, or Portastudio, as pictured in the original user's manual.

had an idea, ten minutes later I could have something on tape. I didn't have to buy a reel or book a studio. I could always find an old cassette somewhere, put tape over the busted-out tabs, and use that." As bassist for the bands Christmas and Yo La Tengo, McNew has since worked in professional studios, but he has often stuck to the 4-Track to make his solo music, released under the name Dump. "When I had an idea for a melody, nothing in the world could have been better," he said. "It has an immediate, urgent quality. It's like, 'I have this music, I have to make it right now and get it to you as fast as possible.' I really responded to that."

Like early rock and roll artists, who worked magic with the limited technical options of their times, many musicians took advantage of the 4-Track's simplicity to find new artistic solutions, concocting their own methods based on the device's basic knobs and effects. As one aficionado wrote in the magazine *Tape Op*, "Inevitably sounds were created on the Porta One that could never be reproduced in a 'real' studio. Were we all crazy, or was there really an intimacy and immediacy being bottled which could never be recaptured?"

A famous answer to that question is Bruce Springsteen's 1982 album *Nebraska*. It was recorded entirely on a 4-Track at a time when the singer-songwriter was a huge star who had crafted many albums in expensive studios. In January of that year, Springsteen said to a member of his road crew, "Go find me a little tape machine." The roadie returned with a Portastudio bought from a local store. With just that device and a pair of microphones, Springsteen recorded an entire album's worth of songs. His team assumed they were demos, but when he tried to re-create the tunes in a proper studio, he realized he liked the 4-Track originals more.

"He kept pulling out that cassette and saying, 'I want it to sound more like this. . . . There's just something about the atmosphere on this tape,'" studio engineer Tony Scott told Tascam. Another engineer, Chuck Plotkin, described Springsteen's re-recording woes to biographer Dave Marsh this way: "The better it sounded, the worse it sounded." That pithy paradox captures the 4-Track's effect on decades of music making. The entire process—centered on the anomalies, quirks, and charms of the cassette tape—created its own sound, blurring the lines between "good" and "bad" sound, "right" and "wrong" techniques. As a tool to make music, the cassette tape had become more than just a format. It was now an aesthetic, a category of artistic creation all its own.

CHAPTER TWO

Creating Music

How Cassettes Helped Launch Movements

To corporations scared of losing control of music, the cassette tape was a killer. But in many places outside of mainstream spotlights, the format actually created music. Offering a cheap, quick way to make, capture, circulate, and share new sounds, the cassette was a crucial conduit for entire scenes, genres, and movements.

In the early 1970s in New York, a new sound percolated at dances in discotheques, parties in parks, and "jams" in underground spots only findable through word of mouth. The common catalyst was the DJ, who spun continuous musical collages with two turntables, a mixer, and a microphone. One particularly innovative DJ, a Jamaican immigrant born Clive Campbell but known as Kool Herc, took small, potent bites of rhythm from records and repeated them by cutting back and forth between two turntables playing the same LP. His "breakbeat" style caught on, and soon more DJs adapted it to their own methods of mixing and matching records on the fly. The resulting sound was called hip-hop, and it spread quickly.

To keep up with the rapid-fire changes happening with New York DJs, you either had to make it to every jam or find some way to hear them after the fact. Enter the hip-hop mixtape, whose origin was somewhat accidental. Anthony Holloway, a.k.a. DJ Hollywood, performed his sets at discos, recording them on 8-track tapes so he could listen back later. It turned out he wasn't the only one who wanted to hear them. "It got to the point where, as soon as I would come outside and say, 'I got tapes!' brothers would roll up and be like, 'Yeah, gimme one of those!'" he told *Cuepoint*. "My tapes would be gone in a flash. That was the real start of the mixtape game."

DJ Hollywood's sets were popular, and they got even bigger when he started talking rhythmically over the music he played (as opposed to flatly announcing songs like a radio DJ) in one of the earliest examples of rapping. "It's not uncommon to hear Hollywood's voice coming from one of the portable tape players carried through the city's streets," reported *Billboard* in 1979. "Tapes of Hollywood's raps are considered valuable commodities." Though Hollywood's innovations got aboveground attention, it was New York's hip-hop underground that really ran with mixtapes, especially when cassettes took over from 8-tracks and reel-to-reels. Fans started taping DJ sets with boom boxes and handheld recorders, either by holding them near speakers or—if they had an in with the performers—hooking them to the venue's soundboard. "You were considered somebody who had the juice if you were able to get a first- or second-generation tape that was really clear," says Davey D, a rapper and historian who grew up in New York in the 1970s. "Mixtapes were the calling card for hip-hop."

D and his cohort collected tapes of the hottest DJs and crews: Cold Crush Brothers, Grandmaster Flash, Afrika Bambaataa, Grandmaster Caz, the Funky Four Plus One. They obsessed over their routines, seeking inspiration for beats and ways to cut records together. Some even used the tapes themselves as sources, dubbing breakbeats and rearranging them to make "pause-button tapes." "You would play the drumbeat and then pause it, and then you would replay it and start it again," D recalls. "The trick was to make your pause-button tape sound like you were actually at a party."

As word circulated and styles mutated, DJs influenced and reacted to each other almost instantly, aided by all the mixtapes flying around New

York. "Cassette tapes used to be our albums before anybody recorded what they called rap records," Afrika Bambaataa told David Toop in the 1984 book *The Rap Attack*. "People started hearing all this rapping coming out of boxes. When they heard the tapes down in the Village, they wanted to know, 'Who's this black DJ who's playing all this rock and new wave up in the Bronx?'" In the oral history *The Art behind the Tape*, Philadelphia-based DJ Jazzy Jeff claims that because of mixtapes, new DJ routines had lifespans of about three parties before that style was nicked and transformed by other DJs.

Things moved so fast, in fact, that once hip-hop entered recording studios and hit radio charts, the underground scene had already moved on. Take "Rapper's Delight," a 1979 song by New Jersey trio the Sugarhill Gang, the first hip-hop single to make the *Billboard* Top 40. "To people outside of New York that was new and exciting, but to people inside that was an outdated style of rap," explains D. "The art form had elevated to another level. Now, how would we know that? Because obviously none of this stuff was on the radio. We knew it from either going to the jams or having those tapes."

Mixtapes were instant cultural currency in New York, but DJs wanted to turn them into actual cash. Sometimes this meant simply making tapes at parties and selling them afterward to fans who wanted to hear it all again or later to people who had missed out. But many DJs discovered a more lucrative market with richer clientele who would pay almost anything to be part of the conversation, literally. These customers— sometimes drug dealers or "hustlers"—would pay DJs to "shout out" their names during sets, then buy a tape from them and blast it around town. It was a way to boast that not only did you have the hottest mixtapes but you were also tight with the DJs themselves. "Mixtapes went hand-in-hand with the drug culture because those hustlers spent so much time in the street," said DJ Jazzy Joyce in *The Art behind the Tape*. Added DJ Clark Kent, "It is more important for the hustlers to have that tape with their shoutout so they can blare it in their car speakers driving down 125th Street, at the car wash on 8th Avenue and 145th, or at Willie Burger where they would post up and park. They wouldn't leave until their shoutout was heard through the speakers so everyone could hear."

Many mixtapes sold for at least twenty dollars, and the most sought-after could fetch over $300 a piece. By the end of the 1970s, prominent DJs could make a decent living just from selling their mixtapes. Grandmaster Flash, who often charged a dollar per minute for tapes ranging from 30 to 120 minutes long, took his game a step further. He made personalized, one-off tapes for specific customers, recorded not at parties but on his decks at home. "The people buying my customized mix tapes were the scramblers, the dealers, people that had money," he told *MTV News*. "I was making a couple thousand dollars a month, easy, doing this." Flash would mix his hottest cuts while frequently shouting out the person who financed his effort. "This is a dedication to my main man Money Mike," Flash says repeatedly on a 1982 mixtape unearthed decades later. "If you have this tape in your possession and you're not Money Mike, you're dead wrong! No copies! No duplicates!"

Hip-hop mixtapes quickly made their way from the streets onto the air. On New York radio stations such as WBLS, WKRS, and WHBI, during shows such as *Rap Attack* and *Zulu Beat*, DJs created audio collages with breakbeats. Accordingly, listeners taped their shows to trade and sell or just to hear themselves shouted out. Many DJs would even give advance warning before giving shoutouts so those getting named could press record. "For listeners, broadcasts were events," wrote John Klaess in his book *Breaks in the Air: The Birth of Rap Radio in New York City*. "They were a forum for delivering shoutouts and performing affiliation. It was a way to get your track heard or your name on tape."

On the strength of their mixtapes, some DJs from parties and clubs became radio jockeys too. One, Harlem-based Kool DJ Red Alert, was possibly the most prolific tape collector in New York. "These tapes were being sold like crazy—I used to make a lot of money off these tapes," Red Alert told *Red Bull Music Academy*. "And I guess that caught [programmers'] ears, because they said: 'Let's get these people who are making these tapes and get them in the studio.'" In 1983, when DJ Afrika Islam started the influential show *Zulu Beat* on WHBI, he invited Red Alert to come on and play his tapes, announcing it weekly as the Red Alert Special. Eventually, Red Alert had his own show on WKRS (a.k.a. KISS FM) as did another prominent New York City live DJ, Chuck Chillout. "We said, 'This is what the streets feel. The same way you hear on the radio, this is what we play in the club,'" Red Alert remembered. "If they don't hear you play that way, they figure you for a phony." As Klaess put it, "KISS's hip-hop DJs composed mixes that chopped and repeated verses, layered new rhymes over old beats, delivered sly commentary on New York's rap scene and social issues, and proceeded rapidly through classic, rare, and fresh-off-the-press breakbeats."

Another major venue for the mixtape was the car tape deck. Beats often boomed from the vehicles of pricey car services that shuttled well-off customers from borough to borough, what Davey D calls "'hood limousines." "These car services knew what was going on in the streets," said Red Alert in *The Art behind the Tape*. "They used to buy tapes from the different crews and DJs and play them in their car while people were traveling." Some of the richest clients would even ask to be driven around for hours without a destination, just so they could listen to the latest tapes. Services such as Godfather and Touch of Class in Harlem and OJ in the Bronx were known to have the hottest mixes. OJ became so big that the company got a shoutout in "Rapper's Delight," and some referred to mixtapes as "OJ tapes."

As hip-hop grew, more key mixtape DJs emerged. Starchild, Lovebug Starski, and Brucie B—who frequently performed at a club called Rooftop with DJ Hollywood rapping over his routines—were self-sufficient entrepreneurs when it came to making and peddling mixtapes. When B had time away from DJing, he would spend it at home dubbing hundreds of mixtapes recorded at Rooftop parties. "While I am rocking in the Rooftop, crews are coming in there from different blocks, and I used to go back in those blocks and make my money, selling tapes," B told writer Troy Smith in 2006. "In one week, I used to make two and three thousand just selling tapes." "I don't care what nobody say—[Brucie B] was the original mixtape king," said Dwight Willacy of Atlantic Records in Justo Faison's 2004 film *The Mixtape Documentary*. "If you had a Rooftop tape, you were the man."

Not all hip-hop DJs were literally men, but few women DJ'd at clubs and parties, and those who did ran up against obstacles. "The guys would always try not to pay me equally, or say my mixtapes wasn't as good as the fellas," said DJ Jazzy Joyce in *The Art behind the Tape*. But Joyce—a "lone soldier for a long time," as she put it—became as influential as any male DJ, and women impacted the scene in other ways too. "DJing is a male-dominated field and it's night based—it takes a lot of sacrifice and a lot of hours," she said in *The Mixtape Documentary*. "Who's going to support that woman who is going out and working late hours? You need some kind of support base, and women are usually the supporters. If you ask the majority of the young men who bought their turntables? It would be their moms."

The hip-hop mixtape's trajectory took another turn with the emergence of Kid Capri, who often made mixtapes at home. "He took it outside of the club," Justo Faison told Davey D in 2005. "It was a crisper sound; there wasn't any background noise from the parties." A master seller, Capri would sit on the sidewalk with boxes full of tapes dubbed the night before, offering them to anyone who walked by. A tape that he originally made for himself, *Old School Vol. 1*, became one of the most famous mixtapes in New York. "Every car that drove by, I heard my name blasting all over the place," Capri said in *The Mixtape Documentary*. "I heard a Mister Softee truck coming down the block playing my tape, then a cop car coming down the block playing my tape, then I started seeing old people with my tape, and then a kid came from the Gulf War, and he said, 'Kid, there were two things that were contraband, the most valuable thing on the ships—cigarettes—and your tapes.'"

Indeed, mixtapes spread outside of New York quickly too. "When they hear these guys rap, they'd make a tape and the tape would travel [to] California, Florida," said Antoine "T. C. D." Lundy in David Toop's *The Rap Attack*. "It just travels everywhere where that person goes. It goes in the army. The guys who used to listen to it recruit into the army and take the tapes with them." The spread wasn't limited to the United States, either. "When I saw that my music was in Japan, I knew it was on!" said DJ Ron G in *The Art behind the Tape*. "I got to Japan and London all from a cassette I made at home."

One crucial hip-hop mixtape scene outside of New York emerged 2,000 miles away. In Houston, Texas, a movement grew around a figure named Robert Earl Davis Jr. Born in 1971, Davis was exposed to music early by his mother, who collected records and made mixes on 8-track tapes, selling them to friends and neighbors. When he was old enough to use a tape recorder, Davis began recording songs off the radio, teaching himself how to pause the tapes at exact moments to catch specific parts of tunes.

At that point, pause tapes were already circulating in Houston. In the late 1970s, Darryl Scott made mixes of R&B and hip-hop using two cassette decks, cutting back and forth so that the beats between songs were continuous. He spent much of high school dubbing his mixes onto blank cassettes and selling them in schools, parks, and anywhere else he could. By the mid-1980s he opened his own store, Blast Records and Tapes, and the teenage Davis became a frequent customer. By then, Davis had figured out DJing on his own, using—and abusing—his mom's albums. "Records he didn't like, he would take a screw and be scratchin' 'em up," his friend Shorty Mac said in Lance Scott Walker's biography

DJ Screw: A Life in Slow Revolution. "So one day he was doin' it and I said, Man, who you think you is—DJ Screw or somethin'?" The name immediately stuck.

Spending countless hours at home practicing his techniques, DJ Screw discovered that slowing albums down changed the sound radically. "I thought the music sounded better like that," he told *Rap Pages* in 1995. "It stuck with me, because you smoking weed listening to music, you can't bob your head to nothing fast." He wasn't the first to figure this out: a Houston resident named Michael Price did something similar when he stuck a screw in a deck to slow its motor down. This result became known as "screwed" music, independent of DJ Screw's moniker, though it would soon become synonymous with him.

DJ Screw's style eventually was called "chopped and screwed," with the first word referring to his habit of cueing up two copies of the same record and cutting back and forth at different points in the songs. His molasses-speed mixtapes became the sound of Houston's streets, aided by codeine-infused beverages known as Syrup or Drank. After a while, he had friends rap over his music—initially to brand his tapes so no one else could take credit for them—and formed a clutch of colleagues called the Screwed Up Click.

In the early 1990s, DJ Screw mixtapes became Houston's hottest hip-hop commodity. Word traveled that you could visit his house—which he bought from tape profits—and buy from him directly. Soon, he had to put up a fence and set selling hours so people would stop knocking on his door all day. "By 7:30 p.m., my whole driveway and street was filled with cars waitin' on me to open my gates," he wrote in *Platinum* magazine in 2000. "And to the rest of the neighbors, they really believed I was conducting illegal transactions. They never saw me come outside." He could garner $15,000 in a single day from cassette sales, enabling him to open his own store, Screwed Up Records and Tapes.

As CDs overcame tapes in the late 1990s, DJ Screw kept his faith in cassettes. As Walker notes, his music pretty much single-handedly kept the format alive in Houston, where it continually blasted from boom boxes and car players. Even after his shocking death at the age of thirty in 2000 (due in part to an overdose of Syrup), tapes remained the DJ Screw currency as his legend grew around the world. "For Screw, dubbing was a tool, the destination for his sound," wrote Walker. "Even though Screw tapes would later end up being released digitally, the origin of their sound was always cassette tape."

At the same time back in New York, the emergence of CDs changed mixtape culture. Where mixtapes were once about the art of DJing, they were now like a buzzy radio station, highlighting new music more than new mixing styles. This was thanks primarily to DJ Clue, whose mixtapes were filled with "exclusives," tracks so fresh they hadn't yet been released or played on the radio. "That changed the game," Faison told Davey D. "Clue used to wait outside the studio for hours for someone to bring him down something new. . . . He would have songs six to eight months before they came out. That set the trend for today." For some older DJs, that trend was unwelcome. "He just played exclusives without any mixing skill," said DJ Jazzy Joyce in *The Mixtape Documentary*. "It was just [saying] his name, and here's the new shit." "What I'm seeing on the mixtapes is you all ain't mixing!" added DJ Hollywood. "You're playing exclusive records and the records you're playing are hot, but it ain't going with the next record you're playing." Kid Capri stopped making mixtapes because of it. "I don't feel like getting exclusive joints and playing them back-to-back is a mixtape," he said in *The Art behind the Tape*. "I do them like stories: a beginning and an end. It isn't just me going all over the place playing the hits."

CD mixes did help the spread of hip-hop, though, and became a source for labels to discover new talent. "Every artist has to participate in the mixtape circuit," said rapper 50 Cent in *The Mixtape Documentary*, "because the majors don't invest in artist development as much as they used to." An entire book could be written about this modern era of mixes, which remain vital into the twenty-first century. But "mixtapes" aren't actually cassette tapes anymore. For decades they came on CD, then moved to digital files traded through the internet or posted as streams on SoundCloud and other websites. They're still called mixtapes because the word "tape" retains underground cachet, implying that the music is fresh, immediate, and unfiltered. "Mixtapes are like me speaking directly to my neighborhood," 50 Cent told *MTV News*. "I can put out a mixtape and just say what I really feel," added rapper Fabolous.

Despite these changes, early hip-hop mixtapes remain crucial documents, preserving history that might be lost if not for cassettes. Listening back to them now, hearing the evolution in beats and cuts and shoutouts, is like audio archaeology. "There are messages upon messages within those tapes," insists Davey D. "I can tell where it was, who was saying what, who the audience was. There are tons of backstories that you can get based on that. That's an important thing within hip-hop."

Just a few hours south from New York down Interstate 95, a sound emerged that was also centered on beats—but in this case, beats from percussionists rather than record-mixing DJs. The musical style that came to be known as go-go began in Washington, DC, in the mid-1970s. Within a decade, it was so prominent that it was essentially a medium of communication for the city's African American community. "Wherever black folk live in the Nation's Capital, you will hear [go-go] groups streaming out of cars as they drive by, from open windows, and—if the headphone wearer has it cranked way up—on the Metro," wrote Kip Lornell and Charles Stevenson Jr. in their 2009 go-go history *The Beat!* "Go-go is more than music. It's a complex expression of our cultural values masquerading in the guise of party music."

Go-go created cultural communication in part because it was an open, participatory form of art. A typical performance would feature extended jams—sometimes just one beat played for long stretches—during which performers and attendees interacted with shouts, gestures, and dancing. This built on conversational musical traditions like gospel music, work songs, and blues jams, updating them for DC's modern sociopolitical currents. "There are three aspects of go-go: the band itself, the lead talker, and the crowd," says Nico Hobson, one of DC's de facto go-go historians. "In order for the band to feel the energy, they need the response from the crowd, and the lead talker connects the two together."

At go-go's height in the 1980s, this conversation happened nearly every day. Groups such as Chuck Brown and the Soul Searchers, Experience Unlimited, and Trouble Funk were just the tip of a wide swath of bands playing multiple times a week all over the city. As with hip-hop, if go-go fans wanted to keep up without spending every free minute at shows, they had to rely on recordings, and those recordings came on cassette tapes. In this scene, they were called "PA tapes," since they usually consisted of exactly what came out of the speakers (i.e., the PA, or public address, system). Many were recorded directly from the soundboard by bands, who passed the master tapes to dealers to sell at shows, on the streets, and in stores.

Hobson himself was a prominent PA tape dealer. He started in the early 1990s after a stint in the military. "I went around trying to obtain [go-go] music, and I didn't like the presentation and the quality of the music I was getting," Hobson recalls. "So it just came to my mind: What is a better way of doing this?" His improvements included detailed documentation (with show dates, locations, and set lists), extensive archiving so he could find specific performances on request, and real-time dubbing that produced better-quality audio than high-speed decks. This meant Hobson often stayed up all night making copies on a rack of twenty recorders.

One particularly novel idea Hobson came up with was turning his street-vendor stand into a replica of the most common listening environment. "I got a car stereo system, a battery which I charged every night, and some Pioneer speakers," he recalls. "When you walked up to my stand, you would get the feel just as if you were sitting in your car listening to it. Because that's how most people listened to the tapes." Hobson recalls four groups whose PA tapes were consistently in high demand: Rare Essence, Junkyard Band, Backyard Band, and Northeast Groovers.

As PA tapes caught on, those in circulation easily outnumbered official releases by go-go acts. Entire stores were devoted to selling them, the most famous being PA Palace, where a customer could read through a log of the store's archive and order any tape they wanted, dubbed right on-site. Individual tapes could cost as much as fifty dollars, and some sold upward of 30,000 copies. Though official go-go releases—mostly twelve-inch singles—did well, the PA tape had an immediacy that studio albums couldn't match. "One of the most popular Rare Essence tapes to date is the 1982 Highland tape," musician Andre Johnson told the *Washington City Paper*. "It was a show where they lost their power and they had to push the percussion for the longest time [while they got] the power back on, and you just hear the percussion going hard. That might not be one of the top PA tapes today if they had taken the errors out."

As in hip-hop, shoutouts also helped make PA tapes hot commodities. It's unclear whether go-go fans actually paid for the privilege, but many would beg bands to call them out onstage, in hopes their names would end up on PA tapes. "You want to feel part of the situation," says Hobson. "When you get your name called out on a tape, you're part of the situation." The demand for shoutouts got so high that Johnson told the *Washington City Paper*, perhaps only half joking, "We do try to play music in between the names." Ultimately, though, the popularity of PA tapes was as much about the way go-go developed as it was about the exclusivity and immediacy of the recordings. Though some mainstream attention arrived in the 1980s (peaking with the success of Experience Unlimited's "Da Butt" single, featured in Spike Lee's 1988 movie *School Daze*), go-go has yet to receive the countrywide recognition that hip-hop garnered. "Once in a while, a go-go record can go national," wrote the *Washington Post* in 1990. "Just don't count on it." "If the world don't want this, fine," said go-go producer Reo Edwards in that same article. "We'll just keep it in DC."

The lack of mainstream success made go-go bands less generous with their master tapes, especially as more record store chains got in on the action and CDs made bootlegging easier and cleaner. Once seen as a necessary evil to get a band heard, PA tapes suddenly seemed like an easy way to get ripped off. "It's hard going into this knowing that when you put a tape out on Tuesday, by Tuesday night, bootleggers have it," Becky Marcus, president of the prominent DC label Liaison Records, told the *Washington City Paper*. "And they're selling it $5 to $10 cheaper than the store." Plus, the sterile, less personal nature of CDs made go-go live recordings seem less special. "I never liked CDs," says Hobson. "I thought it took away from the true nature of the music, for the simple reason that digital doesn't give you the same feel as analog. There's a warmth to the music that can never be duplicated via digital means." Regardless, go-go continues to be a big part of DC, and it wouldn't have spread so far without cassette tapes. "It was one of the arms of the culture," insists Hobson. "PA tapes were essentially the bloodline of that music."

The cassette tape was crucial to another form of dance music, this one born in late-1970s Chicago. One of the first offshoots of disco, house music was centered primarily around a club called the Warehouse and its resident DJ, Frankie Knuckles. He and other DJs mixed disco records with other styles to keep the dance floor moving. At first this was achieved primarily with turntables and mixers, but Knuckles soon began cutting and editing songs into new compositions by splicing together reel-to-reel tapes. At a club called the Music Box, another influential DJ named Ron Hardy used similar techniques. But instead of reel-to-reel tape, Hardy relied on the cheaper and more accessible cassette. And he didn't just use it to premake his sets. Hardy made live, on-the-fly tape edits using the pause and record buttons while music played on his turntable, then mixed that tape in with the original tune. This helped spark a trend in house music in which DJs didn't just play records but became remixers, producers, and even composers.

As Micah Salkind points out in his book *Do You Remember House?*, the kind of mixing that Hardy, Knuckles, and other DJs performed went deeper than matching beats. They would take two disparate records and juxtapose them, isolate small portions of songs to create danceable loops, change speeds of music the crowd already knew, add documentary-like sound effects (Hardy often panned a train sound across the dance floor), and mix in aural iconography like Martin Luther King Jr. speeches to create cultural commentary. In Hardy's case, audience response turned him into a living legend. "I'd never seen anyone yell for a DJ before Ron Hardy," DJ Pierre told *Red Bull Music Academy* in 2015. "I mean, they were screaming his name. People were so passionate that they would start crying."

Cassettes also helped DJs and fans become promoters of house music. As with hip-hop and go-go, house music circulated on tapes DJs made themselves. Hardy often sold his from the booth for twenty dollars apiece. In this way, he could disseminate his work in real time. Sometimes an artist would make a new track on cassette and hand it to Hardy that day, later hearing it in his set that night. "Hardy acted as an informal A&R rep for the underground," wrote Salkind, "using his platform to disburse cultural capital laterally across Chicago's house scene."

When the cassette tape became popular, heavy metal had already been around awhile, pioneered in the late 1960s by bands such as Led Zeppelin, Black Sabbath, and Deep Purple. But within a decade, some of metal's more devout practitioners and aficionados were tired of the bloated, arena-rock direction that popular metal had taken and looked outside the mainstream for new sounds. It turned out that many could be found only on cassette tapes.

In the late 1970s, one of the most exciting metal tapes to emerge came from British band Iron Maiden. Working quickly and cheaply in a studio in 1978, they recorded four raw, fast tunes. They then dubbed them onto cassette tapes and passed copies around to people they knew, including Neil Kay, a DJ at his own London club, the Bandwagon Heavy Metal Soundhouse. In Mick Wall's Iron Maiden biography *Run to the Hills*, Kay called their tape "the most impressive demo I've ever had delivered to me." Whatever songs he played the most at Soundhouse were listed in music weekly *Sounds*, and by spring of 1979, Iron Maiden's "Prowler" hit number one on that chart. "All these fans [were] asking where they could buy one of our records, and when we told 'em there wasn't any yet they couldn't believe it," said songwriter Steve Harris in *Run to the Hills*. "A lot of 'em just assumed we must already have a record deal of some kind, but we didn't . . . so then, they'd be, like, 'Well, where can we get a copy of the tape?'"

Iron Maiden eventually pressed the recordings onto a vinyl release called *The Soundhouse Tapes* and signed to a major label in late 1979. But the tape had already circulated around the world, as *Sounds* labeled Iron Maiden and other new bands the New Wave of British Heavy Metal (NWOBHM). Many of these groups had few official releases on commercial labels, so demos and live bootlegs were the main way to hear them. But since those tapes weren't sold in stores, curious listeners had to rely on person-to-person, mail-based interactions. In other words, NWOBHM fans needed a way to find one another.

Luckily, magazines—particularly low-budget "fanzines"—started popping up, with names such as *Metal Fury, Metal Mania, Metal Rendezvous,* and *New Heavy Metal Revue.* Many included classified-ad-style sections where fans could list contact info, what bands they liked, and what tapes they were looking for. "Crazy, deaf, metal maniac seeks same to trade demos, albums, etc., and all other info on metal from all over the world," wrote one reader in *Metal Forces'* Penbangers section, before listing ten groups he loved. "There must be someone in the Sheffield area who suffers from lack of, or no, Metal friends for gigs, friendship, etc." wrote another to the Penpals section of *Kerrang!* (an offshoot of *Sounds*). "Please write to this desperate 16-year-old Indian."

"When I got my letter published in *Kerrang!*, that was about the best thing that happened to me," Northern California–based metalhead Ron Quintana told *Metalcore.* "Literally 100 American music traders got a hold of [me], and another 100 European traders. Every day was like Christmas." Down the coast in Southern California, Quintana's future friend Brian Slagel had a similar experience. "Whatever bands I was a fan of, I wanted to have everything by them, so I put ads in magazines," he says. "Then every day I would come home from school, hoping to find a package in the mailbox." One such package included an AC/DC live tape to which Slagel's tape-trader friend had appended three songs from Iron Maiden's *Soundhouse Tapes.* "I started freaking out," Slagel wrote later. "The music was so exciting. It sounded like the future. My future." Across the country in suburban Maryland, Jim Powell sat waiting for tapes too. "In one of my classes in high school, there was a phone in the back of the room," he remembers. "You weren't supposed to use it, but if I ever saw an opportunity, I would call home and ask my mom if any packages came in the mail. My everyday life was listening to tapes, making tapes, and trading tapes."

These metal fans collected demos, rehearsals, concerts (sometimes recorded by the traders themselves), radio shows, and more. Often, each iteration of any traded tape would get modified and personalized, not just with handmade art and track listings but with extra music if there was room left. Someone asking for a fifty-minute Iron Maiden show might get a sixty-minute tape with other bands at the end, and maybe those bands suddenly had a new fan. As metal proliferated through these interactions, the community came to feel like a secret club, limited only by whatever spare money and time fans had to dub and mail. "When you would meet someone into these same kinds of bands, you would immediately draw a connection with that person," says Powell, "because the scene was very small at that time."

Powell, Quintana, and Slagel all started their own fanzines and eventually put out records, booked shows, ran labels, and managed groups. Quintana's and Slagel's influence on metal started early, when they each met drummer Lars Ulrich, a fellow metalhead and tape trader. Inspired by NWOBHM bands, Ulrich put an ad in Los Angeles magazine the *Recycler* seeking musicians who liked Iron Maiden. Singer and guitarist James Hetfield responded, and Ulrich named their project Metallica, borrowing the moniker from Quintana (who considered it for his fanzine before choosing *Metal Mania*). Before he had even filled out the Metallica lineup, Ulrich heard Slagel was putting out a compilation called *Metal Massacre* and convinced his friend to let Metallica participate (though Slagel tested that friendship a bit when he mistakenly spelled their name with two *t*'s on the back cover).

Soon after *Metal Massacre* came out, Metallica recorded seven tracks intended for release on a punk rock label, which backed out once they realized Metallica was a metal band (though their high-speed "thrash" metal wasn't that far from punk). Ulrich suggested the band instead dub the songs to cassette and send copies to fanzines and traders. Though this was done out of financial necessity, Ulrich also knew underground word of mouth could be as powerful as any record label's reach. "[Fanzines] were just going nuts over Metallica," Patrick Scott, who did most of the dubbing and mailing, told Mick Wall in the Metallica biography *Enter Night*. "Even in countries where we thought the cool bands were, they thought Metallica was the coolest band. It was a fun time, running to the mailbox every day." "There were so few ways for people to discover bands that if it wasn't for that tape trading network, you wouldn't have had that buzz," recalls Slagel. "Lars worked that system perfectly, getting

METALLICA.DEMO.

1. HIT THE LIGHTS
2. THE MECHANIX
3. MOTORBREATH
4. SEEK x DESTROY
5. METAL MILITIA
6. JUMP IN THE FIRE
7. PHANTOM LORD.

Brian Slagel's original copy of the 1982
Metallica demo known as *No Life 'til
Leather*. (Photo by Brian Slagel)

this tape out to everybody and getting this tremendous buzz going." Within a few years, Metallica released records on independent labels, got signed to a major, and eventually became one of the biggest bands in the world.

While Metallica ascended, the metal underground kept evolving, and cassette tapes remained key. In the mid-1980s, two related styles— death metal and grindcore—circulated primarily through tape trading. In Virginia, a teenager named King Fowley learned about trading from Jim Powell, diving headfirst into bands such as Anthrax and Slayer. Not long after, he formed the band Deceased and started sending around demo tapes. "We were just like, 'Send us some stamps, whatever, and we will mail it,'" Fowley recalled in Jason Netherton's book *Extremity Retained: Notes from the Death Metal Underground*. "We made a lot of friends that way, and I discovered a lot of bands from all over the world." "It was a super organic thing; it was just one friend suggesting to another friend," says Albert Mudrian, editor of *Decibel* magazine. "If you wanted to hear more death metal, the only way you could do it was through these bands trading tapes and getting ideas off each other. For a long time, there were only, like, two death metal bands that had records. You were listening to maybe a maximum of four real records, and then everything else was on tape."

The underground network for death metal and grindcore quickly became international, connecting bands such as Napalm Death and Carcass in England, Death and Morbid Angel in America, and Nihilist and Carnage in Sweden. In some countries, the mail was so controlled that visitors from the West had to bring tapes with them, as if the cassette were contraband. "The censorship culture in Czechoslovakia back then in the late 1980s was very strong," said Bruno Kovarik of the band Hypnos in *Extremity Retained*. "After 1989, when the communist regime finally broke down, the borders were opened up, and we could send letters and packages to the whole world."

Along with the stealth nature of tape trading, there was the feeling that the music itself was unfiltered and unsanctioned. "I don't think any big producer in the thrash metal days would have allowed that kind of guitar sound to be recorded," said Dan Swanö of the Swedish band Edge of Sanity. "[But] the bands would say, 'This is what our guitars should sound like,' and [the recordist] would say, 'It sounds like a chainsaw but who cares? Fine with me.'" The rawness of the music was thrilling for fans but also helped artists grow outside of a pressurized spotlight. "These bands would go through, like, four, five years of demos and rehearsals [before getting signed]," Mudrian says. "So they actually had time to hone their craft. By the time they get to a debut album, the material's really good. Part of what really hurt the scene in the nineties is so many bands were getting scooped up after putting out one demo, and then they put out a record, and that was it. Some bands didn't have time to develop."

Metal tapes faded in the 1990s as the CD format rose, and some traders became more interested in collecting established bands and spreading rarities by those groups rather than finding the raw material of unheard newcomers. Yet some metal scenes have stuck with the cassette. Black metal in particular uses cassette tapes to stay underground. Demos and rehearsal tapes are passed through the mail or sold in small shops, featuring bands such as Darkthrone, Emperor, and Dead Reptile Shrine. As Edwin Pouncey wrote of the latter in the *Wire*, "There is an almost talismanic feel to their cassettes that convincingly evokes a presence of something spectral joining in—coupled to a sensation of edgy excitement as the group's rustling and chanting produces sounds that, to the awestruck listener, seem intoxicatingly forbidden."

When punk rock emerged in the 1970s, cassettes immediately played a role. Take Television's first demo tape, recorded by Brian Eno. It circulated for what seemed like an eternity before the band finally got a label contract. But most early punk bands—from protopunks the Stooges to New York originals Ramones and UK snots Sex Pistols—signed to major labels quickly, and their music came out on both vinyl and tape. Cassettes were more important to a scene that grew out of punk in the late 1970s and early '80s: indie rock. Though the term came to mean many things over the ensuing decades, at first it signified rock music that was made, released, and distributed independently of the corporate music industry.

Many indie rock labels of this era released music both on LP and tape, and eventually CD. But a significant subset preferred to go primarily, or even exclusively, with tapes. The first prominent cassette-only label in this scene was independent both out of necessity and attitude. In the late 1970s, Neil Cooper, a former talent agent for major label MCA, booked bands at The '80s, a nightclub on New York's Upper East Side. He started recording concerts there, hoping to release them himself. But many of the performers were already under contract with labels, and those who weren't often aspired to nab similar deals.

Cooper convinced some bands that cassettes would be a good way to release "works in progress, or stuff that they didn't think was good enough for an LP and a major-label commitment," as he told *Billboard*. Some label contracts covered only vinyl releases, so cassettes were fair game. In 1981, Cooper launched Reachout International Records (a.k.a. ROIR, which he insisted be pronounced "roar"), telling the *New York Times*, "There are plenty of advantages to cassettes. If you become bored with the music on them, you can record something else over it." His first release was *Live in New York* by James Chance and the Contortions, one of the stalwarts of a noisy downtown scene that became known as no wave. "I'd just as soon listen to a cassette as an LP record," wrote downtown denizen Glenn O'Brien in the liner notes. "Cassettes don't scratch and they last longer. . . . Also you will be able to play this while roller skating, hoeing in the garden, motoring in the country, etc."

Cooper quickly began churning out cassettes by his favorite New York artists—Richard Hell, Suicide, New York Dolls—and distributed them himself to stores. Just a year into ROIR's existence, Cooper landed a semi-hit: the first album by groundbreaking Rastafarian punks Bad

A print advertisement for New York's ROIR,
boasting its status as one of the only labels
releasing cassette tapes exclusively.

Brains, which sold over 60,000 copies on cassette. "This tape was a fucking game-changer," Ian MacKaye of hardcore punk band Minor Threat
would recall years later. "They managed to get it into important record
stores . . . and it was also just unimpeachable." Cooper also released
numerous compilations of New York–based music, touting himself as
a scene archivist in the spirit of folk documentarian Alan Lomax. A
lofty comparison, perhaps, but ROIR compilations have held up well as
artifacts of an era that flew by fast. Many of the bands didn't last long,
and without ROIR, their music might have vanished with them. Cooper
eventually released some CDs, but the reason he started a cassette-only
label stuck with him forever. "We're one of the few remaining totally
indie labels in the U.S., which is almost impossible," he told *Billboard*
in 1999. "We have no outside income other than what we beg, borrow,
and steal."

On the other side of the country, Cooper's interest in local artists was
echoed in the fandom of Bruce Pavitt. Enrolling at Evergreen State College in Olympia, Washington, in 1979, he discovered the campus radio

station KAOS and its music director John Foster, who spearheaded a magazine there called *OP*. Pavitt soon got his own show, dubbed *Subterranean Pop*, and by the next spring started his own magazine of the same name (later shortened to *Sub Pop*). It featured reviews of independent releases grouped by regions of the United States, reflecting what Pavitt calls the "tribal energy" of small scenes. "Only by supporting new ideas by local artists, bands, and record labels can the U.S. expect any kind of dynamic social/cultural change in the 1980s," Pavitt wrote in the first issue of *Subterranean Pop*. "We need diverse, regionalized, localized approaches to all forms of art, music, and politics. . . . Tomorrow's pop is being realized today on small, decentralized record labels that are interested in taking risks, not making money."

A few issues into *Subterranean Pop*, Pavitt decided to spread this art via cassette tape. Inspired by Australian tape publication *Fast Forward*, Pavitt made his fifth edition a cassette compilation rather than a print magazine, aided by fellow KAOS DJ Calvin Johnson. "Most compilations at that time were regional compilations, with all hardcore bands from Boston or something like that," says Pavitt. "What I very consciously wanted to do was transregional compilations, so people in different scenes could hear what was going on in other ones." Released in 1981, *Sub Pop 5* featured artists from Washington, Kansas, Virginia, Wisconsin, and Missouri. Pavitt's "transregional" idea was mirrored in England, where the music weekly *NME* released *C81*, a tape collection of mostly UK-based post-punk bands that was as influential as *Sub Pop*. Its success spurred the magazine to make thirty-six tape compilations over the next seven years, including the equally important 1986 edition *C86*.

As with the print version, cassette editions of *Sub Pop* were made cheaply, using blank tapes from a nearby wholesaler and the dubbing services of a local technician. "We could do runs of ten, twenty, however many, whereas with vinyl you'd typically have to do a minimum run of 500," Pavitt remembers. "Once we got some tapes, we would cut up little zines and staple them and stick them in there." At five dollars apiece—more expensive than a single but less than a full-length vinyl album—*Sub Pop 5* sold over 2,000 copies, most of which were bought directly from Pavitt through the mail. The revenue went right back into future editions of *Sub Pop*.

Inspired by his work with Pavitt, Calvin Johnson started a cassette label of his own. After inviting the Olympia band Supreme Cool Beings to play live on his KAOS show in 1982, Johnson wanted to release a recording of their set but couldn't afford to press it on vinyl. So he put a tape together step by step, buying 150 blanks from an Olympia supplier, paying his friend Pat Baum to dub them on her one-copy-at-a-time setup, making photocopies of hand-drawn art by Supreme Cool Beings' Heather Lewis, and adding a hand-drawn *K* with a shield scrawled around it, christening his new label. This first K release was rejected by all the distributors Johnson sent it to, so he carted it around from store to store himself, which he later described in Michael Azerrad's book *Our Band Could Be Your Life* as "the most exciting thing that had ever happened in my life."

Johnson followed up K's debut with cassette compilations of unpolished music housed in simple packaging. Two of those compilations included Beat Happening, a band Johnson started with Lewis and Bret Lunsford. Their music would soon become, along with K Records itself, one of the most well-known symbols of 1980s do-it-yourself indie rock, with an ethic that valued community and frugality over corporate-backed, PR-driven machinery. "A cassette is great for a local scene like Olympia because a band can release a cassette and not have to spend their would-be savings," Johnson told *Flipside* in 1986. "You make up as many as you need, they're cheap, and if you don't sell them, you just reuse them." "When Calvin started K, the medium was part of what he was promoting as much as the music," Lunsford said in Mark Baumgarten's book about the label. "Hey, now everyone has the equipment to stage this revolution against the corporate ogre, so buy some blank tapes, record yourself in your living room, release it!"

K has since put out music on other formats, but the label always maintained an independent spirit, evidenced in its International Pop Underground series of seven-inch singles and multiband festivals. Pavitt made a similar move in 1986, releasing a vinyl compilation called *Sub Pop 100* that followed the transregional philosophy of his earlier cassette releases, and turned Sub Pop into a full-fledged label. Moving to Seattle, he tapped into the city's nascent grunge scene, elevating his imprint—thanks in part to the surprising success of Nirvana—into one of the most successful and famous independent rock labels ever.

As well as being icons of indie rock, K Records and Beat Happening would also become associated with a subgenre known as "lo-fi." The name is short for "low fidelity," with "fidelity" indicating how closely a recording matches the original sound it reproduces. If professionally recorded sound was high fidelity, lo-fi stood in opposition, branding music that for whatever reason didn't hew to conventional technical standards. Lo-fi musicians often recorded on their own, without producers or label budgets, in their own spaces, on cassette. The hissy, distorted sound of tape was the price they paid for working this way. But many also treated that as a plus, adding texture, mystery, authenticity, or personality to their work.

Many years before people used the term "lo-fi," an artist in Nashville, Tennessee, was already living it. Born in 1952, R. Stevie Moore was immediately surrounded by music. His dad was a session musician who backed up luminaries such as Elvis Presley and Roy Orbison and got Stevie a singing gig on a professional recording when he was just seven years old. But soon, Moore rebelled toward his own way of making music. "I always hated that whole country scene and the music itself," Moore told the *New York Press* in 1998. "I would go home from these sessions and do some crazy, under-the-headphones, home recording. I became a hero in my own head, in my own room."

Moore quickly became a prolific home recorder on reel-to-reel tapes, composing and recording hundreds of songs in all sorts of styles. His own high school newspaper noted his obsession, reporting that "this blonde, introverted B-plus student has a cool new hobby of playin' around with his Tape Recorder, and he's started to compile songs 'n sounds with his friends." Moore put out some vinyl releases in the late 1970s, but his cache of recordings became so large that there was only one affordable way to make them available: on cassette.

In 1982, "fueled by an almost fanatical determination to make [my] music available before [I die] young," he launched the R. Stevie Moore Cassette Club. Members received photocopied lists of titles, each of which included a "listenability quotient" to indicate how lo-fi it sounded. He also offered ninety-minute custom-made tapes with selections from his vaults for eight dollars apiece (with a money-back guarantee). Though Moore had taught himself to record with decent quality on his home equipment, his tapes were still stylistically idiosyncratic. He would often leave in mistakes, include fragments of conversation, and make abrupt cuts from song to song, as if flipping around his own internal radio dial. "My cassettes are a diary of sound, a very personal kind of thing," he told the *New York Times* in 1987. "This is what I do, writing songs and building soundscapes. It's almost a kind of sickness."

Considered by some to be the godfather of lo-fi, Moore has some-times eschewed that description, insisting that his prolific output was simply about exploration. Home recording and self-releasing on cassette just happened to be the easiest and quickest way to do it. But as recently as 2011, he touted cassette tape as a valid document. "I'm sick of people even talking about demos—all of my work is demos," he told *Tape Op*. "They're recordings. It doesn't mean they need to be perfected with sheen, polish and reverb. I've gone through the whole thing of trying to remake some of my home recordings, and there was something lacking. They *sounded* amazing, but they were forced and they weren't inspired. It always takes away."

Not long after Moore began cranking out tapes, another prolific home recorder started doing the same, tucked away in his parents' basement. Growing up in the 1960s in West Virginia, Daniel Johnston was equally obsessed with the Beatles and movies. "When I was little, I pounded on the piano and pretended I was making soundtracks to horror films, monster movies," he told *Time Out* in 2013. "I spent the day doing that." As a teenager, he acquired a Sanyo recorder for fifty-nine dollars, bought some blank tapes from RadioShack, propped his gear on top of the family piano, and hit record. He gave the resulting tapes out to family and friends. If he didn't have a way to dub from tape to tape, he would actually reperform his songs on a new cassette for each recipient.

These initial tapes, which Johnston gave names such as *Songs of Pain* and *Don't Be Scared*, contained bracingly honest, open songs, their intimacy and immediacy enhanced by cassette. With all the tape hiss, room noise, incidental sounds, and rawness of Johnston's voice and playing, the music was both achingly personal and knowingly funny. He would even include snippets of audio vérité recorded around his house, often featuring him talking or arguing with his parents. On one tape, his mother says, "We have enough depressing songs. I'd like to see you do something fun." "I don't mean to be depressing," Johnston responds. "Sometimes it just turns out that way."

Throughout the early 1980s, Johnston continually released cassettes, featuring his hand-drawn artwork of surreal creatures, mutated boxers, and figures resembling one of Johnston's favorite characters, Casper the Friendly Ghost. Living in Austin, Texas, he hung out on the street and handed out tapes to passersby; later he got a job at McDonald's and stuck tapes into bags of burgers that he sold. By the time he gave his first public performance, many people in Austin knew his music from

Daniel Johnston's early cassette tape releases, some of which he sold on the street and gave away at his job at McDonald's. (Photo by Jeff Tartakov)

his tapes, and he scored an appearance in an MTV report about local music. But he didn't abandon the cassette format. "He was a big believer in cassettes throughout the '80s," says friend and eventual manager Jeff Tartakov. "He viewed each cassette he recorded as an album, and there was never any consideration of finding someone to release the cassettes on vinyl until after his MTV appearances."

Tartakov had started his own tape label in Austin called Stress Records and wanted to help Johnston spread his songs, which already numbered over 200. "He was spending every dollar he made at McDonald's on the cassettes he was giving away," Tartakov remembers. "So in early 1986, I offered to take over the cassette dubbing after convincing him that he should at the very least be breaking even." Tartakov got Johnston's cassettes to stores and distributors around the country, and opportunities poured in. Early Johnston tapes were reissued on vinyl, and he ventured to New York in 1990 to record his first studio record with renowned indie producer Kramer and befriended one of the biggest indie rock bands around, Sonic Youth. Johnston eventually signed with major label Atlantic and became world famous, thanks in part to Jeff Feuerzeig's documentary *The Devil and Daniel Johnston*. But his music retained the purity and honesty of his earliest work, and he would always be identified with lo-fi.

Johnston's years of renown were filled with ups and downs, as issues with manic depression led to both personal and professional turmoil. Throughout it all, and since Johnston's passing in 2019, Tartakov has kept his cassette catalog alive. "I felt strongest about the importance of the cassettes in the late 1990s after Daniel had been dropped by Atlantic and was struggling to move forward," he says. "Keeping the cassettes available during those years led to many bands continuing to discover and cover his songs."

For one lo-fi musician, discovering Johnston's tapes was a personal revelation. In the mid-1980s, Lou Barlow played bass for Massachusetts band Dinosaur Jr., which at one point was touring with Sonic Youth. "We were standing outside after a show in Buffalo and [Sonic Youth] played Daniel Johnston out of their huge boom box," he recalls. "I was like, 'Oh my God, this is just like what I've been doing!'" Growing up in Michigan, Barlow latched on to cassettes before he even started making music. His dad would craft audio letters for relatives, mixing his messages with recordings of dinner parties or his children's voices. Inspired, Barlow committed his own routines to tape. "I took it upon myself to find any way to use a combination of buttons on the cassette player to make the craziest sounds," he remembers. "Like cartoonish voices or slow voices or jamming pencils and shit into the gears."

Moving to Massachusetts as a teen, Barlow sent tapes to a friend back in Michigan, filling them with songs, skits, and snippets recorded off the radio. At first, he assembled a crude recording system with two tape decks and some cords he bought at RadioShack; eventually, he saved up enough money to buy a Portastudio 4-Track. "I got really good at it," he says. "I felt this sense of mission and purpose to really fuck with these tape recorders and to ultimately work the collage aspect into my songs. It would be like these super-lo-fi, fucked-up, psychedelic tape song clashes."

In the meantime, Barlow joined Dinosaur Jr., and when it came time to record their second album, 1987's *You're Living All Over Me*, one of his tape collages was included at the end. "Poledo" marked the public debut of Barlow's solo project, which he would soon call Sentridoh. He also recorded a tape called *Weed Forestin'* filled with intimate, hissy folk songs and gave it away at stores with copies of *You're Living All Over Me*. "I've made all my tapes for myself, but I'd go crazy if I kept it to myself," he told *Rockpool* in 1990. "I figure if it's in me and I'm thinking about it, then someone else is thinking about it too, so it's pretty necessary to put this stuff out."

Releasing Sentridoh material ended up being as valuable socially as it was artistically. "Those tapes were literally how I made friends, the friends that I have to this day," Barlow says. "When I met people who actually liked it, it was like, 'Oh, okay, you're my friend for life now . . . and we will soon be spending days and days together in a van.'" That literally happened with Eric Gaffney, who had been making tape experiments of his own when he met Barlow. The two combined forces under the name Sebadoh to make a cassette called *The Freed Man*, which they sold for a dollar at stores.

As Dinosaur Jr. and Sebadoh rose in stature, Barlow continued to release cassettes of his home-recorded music, both on his own and through small indie labels. In one interview, Barlow mentioned that if anyone wanted to release his work on tape, he had a lot ready to go. Upland, California, resident Dennis Callaci responded and in 1991 released the Sentridoh tape *Losers* on his fledgling cassette label, Shrimper. A music fanatic who self-published a zine and played in a band, Callaci started Shrimper while working at a record store. "One of the guys that owned half of the store had a cutout company," says Callaci, referring to older, "remaindered" releases that were sold cheaper than originally priced. "He had these tapes that were a quarter a pop, so I bought them, taped over them, and just threw a sticker on them." Because of bleed through, on some Shrimper tapes you can hear ghosts of the music Callaci was erasing.

Along with Barlow's work, many of the releases on Shrimper were by artists who would help define lo-fi: Franklin Bruno's Nothing Painted Blue, John Darnielle's Mountain Goats, and Callaci's own band Refrigerator. Their DIY approach was mirrored in Callaci's process. He would stay up all night dubbing copies of Shrimper releases at home one at a

A hand-dubbed tape of Sebadoh music
made by Lou Barlow for the author, 1991.
(Photo by Marc Masters)

time. "I just like the ritual of it," he says. "You're tethered to this machine, and every eight to twelve minutes you have to flip the tape. It reminds me of working graveyard shifts at the gas station."

That ritual was likely as important to Charlie McAlister, who began releasing tapes of his own music in the mid-1980s, first under a variety of humorous monikers (Compost Heap, Flux Zebra, Head Injuries), then simply as C. McAlister. Born in South Carolina in 1969, he started recording his own songs in high school, eventually appearing on over 100 releases. He hand-drew and photocopied his own covers (and would become a respected, gallery-worthy visual artist) and traded tapes through the mail constantly. One of his very first cassettes, *Bad Music for Good People*, includes the note "send in your bad music for inclusion on a future tape." His work, much like that of R. Stevie Moore and Daniel Johnston, was a mix of primitive rock, folk, and other styles, filled with odd instrumentation and bearing titles to match: *Florescent Chicken Package Deal*, *The Fake Puntegg Roll Bomb Pass*, and *Tire Iron Bloatee at the Motorcycle Race*.

Swaths of other independent artists got their starts—and built their reputations—with lo-fi cassettes. In the early 1980s, longtime home recorder Robert Pollard put reams of songs on tape under the name Guided by Voices, at first just using a condenser mic and a stereo, then moving to 4-Tracks. In the late 1980s, the band Silver Jews began when

David Berman and his friends Stephen Malkmus and Bob Nastanovich (both of the group Pavement) lived together in a basement apartment in Hoboken, New Jersey, recording songs onto a cheap tape recorder placed atop their TV. They would "mix" by each standing at different distances from the deck. In the early 1990s, Liz Phair started "making up crazy little songs in my bedroom," as she told *Jezebel* in 2018, crafting three cassettes under the name Girly-Sound. Those tapes zipped around the underground so quickly that she soon got a record deal and eventually signed to a major label. "I never looked at Girly-Sound as the precursor of anything," Phair said. "It was just the way to remember songs and put them down and then forget about them."

As the end of the 1990s approached, CDs overtook cassettes, and digital recording became easier and cheaper. But lo-fi remained relevant as an aesthetic, a way to convey authenticity and directness rather than polishing music in glossy studios. That idea might seem a bit strange to people who used cassettes because it was all they could afford. "Aesthetics? I didn't even know what those were," insists Callaci. "I was doing it so quickly—the only real thought I had was that I had to capture stuff that was not going to be documented otherwise."

For Lou Barlow, though, aesthetics matter. "I'm such a huge champion of lo-fi," he says. "I think it's been amazing how lo-fi is part of the whole arsenal that people use to make music now. If I was a part of that in some way, that's fucking great." In 2019, when digging into a stash of old tapes for a series of releases on the label Joyful Noise, Barlow was inspired anew. "One thing I love about my early 4-Track recordings is the way that my strumming was captured," he told Vish Khanna on the *Kreative Kontrol* podcast. "It seemed to really work, and I had just totally lost that for decades." As a result, Barlow made his next album, *Reason to Live*, using a cassette player he found at a yard sale. Lo-fi, direct, and captivating, it evokes his early work while reflecting the changes his music went through in the ensuing decades. "I was like, 'Fuck it, I want to actually capture the way I play . . . make it sound the way I want it to sound,'" he said to Khanna. "So I was like, 'I've got to get cassettes.'"

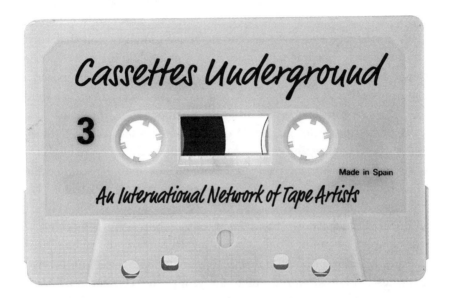

CHAPTER THREE

Cassettes Underground

3

An International Network of Tape Artists

Made in Spain

Ask anyone who worked with John Foster at KAOS, and they'll say he is a visionary. In the mid-1970s, as music director at Evergreen College's radio station, Foster was obsessed with musicians and labels operating outside the rarefied network of major corporations. So he came up with a new rule: 80 percent of the music DJs played on KAOS had to be on independent labels. In ensuing decades, stations across America adopted similar formats, distinguishing college radio as a reliable source for alternative music.

"I was frustrated that there were a lot of these small labels that weren't getting played," says Foster. "I wanted to be an evangelist for this whole idea." To that end, he started the Lost Music Network (LMN) and in 1979 began publishing a magazine called *OP*, titling it after the next two letters in the alphabet. At first, *OP* was an insert in KAOS's program guide, but it soon became a stand-alone publication covering all kinds of artists and music, with a plan of ending after twenty-six issues (one for each letter of the alphabet). "We weren't trying to make a buck; we weren't trying to make a certain style of music happen," says David Rauh, who sold advertising space for *OP*. "We were just trying to see if it would help people be more in touch with each other about stuff they were doing."

A few years in, Rauh and Foster noticed that a lot of the music sent to *OP* came on cassette tape, often hand-dubbed and hand-decorated. They also recognized that, despite the conventional notion that vinyl albums were official and tapes were merely demos, these artists saw the cassette as a worthy format, not a stepping stone. So Rauh proposed a column in *OP* that covered only tapes, and Foster was game. "We were like, 'Does anyone give a shit?'" Rauh remembers. "'Let's find out!'"

It turned out a lot of people did. In the summer of 1981, Rauh launched his new section called Castanets (because it sounded like "cassettes"). Wary of looking conflicted as an ad salesman, he adopted the pseudonym Graham Ingels after his favorite comic book illustrator. "We literally got bushel baskets of cassettes," he says. "I had assumed I'd just sit down with what we got and listen to it all, but within a month that was already way out of reach." "It was a culture ready to catch on fire," adds Robin James, whom Rauh enlisted to help write the column's brief, basic descriptions of tapes, which always included contact info. "Most people wanted to hear the stuff for themselves," explains Rauh. "And we were just giving them a road map." It was an approach similar to that of a number of small publications at the time, including *ND*, *Factsheet Five*, and *UnSound*. Foster and Rauh reacted to a growing phenomenon, but they also helped accelerate that growth. *OP* became, as Neil Strauss insisted in Robin James's book *Cassette Mythos*, "the single most important influence on home tapers."

This global community of home tapers was ignited by the affordability and convenience of the cassette tape, much like tape obsessives in hip-hop, go-go, metal, and indie rock. But this underground cassette network was much more far-flung, a seemingly endless web of interpersonal connections, collaborations, and affiliations made primarily through the mail. Its participants were linked not by a regional scene, or a developing sound, or a musical movement. They connected through the cassette itself, with which they made and exchanged their own self-propelled, self-controlled art, blurring categories and dodging commercial pressures. "You had folk musicians, academics, tape collage people, noise people, industrial people, punks," says GX Jupitter-Larsen, who began making music in the late 1970s under the name the Haters. "It was a little bit of everything all mixed together."

Many of these home tapers got inspiration from the UK group Throbbing Gristle. In the early 1970s, some of their members were part of COUM Transmissions, a collective that often sent and received art through the mail. After forming Throbbing Gristle, they made two tapes, *The Best of Vol. 1* and *The Best of Vol. 2*, dubbing them themselves and giving them to friends. Their first distributed release, 1977's *The Second Annual Report*, was partially recorded in a practice space on a Sony cassette recorder. They also put out numerous cassettes on their label, Industrial, while recording most of their shows on tape (some of which were released as a set of twenty-six in a suitcase). Most important, their newsletter requested that readers send them tapes, an early catalyst for mail-based cassette networks.

In the scene that *OP* covered, the postal system was crucial. Tape trading was in part an offshoot of the phenomenon known as mail art, wherein creators would share, collaborate, and disseminate visual and text-based art via the post office. "You would dub a cassette, hand-make a cover, and just send it to someone," remembers Jupitter-Larsen. "It wasn't a release the way most people would think of a release. It was kind of a personal audio letter, in an edition of one." As a result, mail art fostered creative communities. "Artist types want to be loved," wrote Italian mail artist and taper Vittore Baroni. "The warm direct contact with hundreds of allied individuals worldwide was much more satisfying than the occasional feedback from a few viewers or critics at art exhibitions." "The tape community was tiny compared to today's online network and it felt intimate," prolific taper Zan Hoffman told fellow artist Don Campau. "It took a concerted effort to be part of this and accordingly there was a sense that 'we were all in this together.'"

Often, the presentation of these homemade cassettes was as integral to the art as the music itself. Since the postal system accepted almost anything with an address and postage, artists created their own packaging by, for example, buying toy dolls at thrift stores, cutting holes in them, and sticking tapes inside. "It became kind of a contest," says Robin James, who made his own music on tape. "When you got a tape from Das [an artist in San Francisco], you'd also get a playing card and slides and little toy soldiers and stuff from inside Cracker Jack boxes." James's 1985 compilation *The Box* housed three tapes in a wooden box filled with cedar boughs, which inadvertently disintegrated and got lodged in the cassettes. Campau sent his 1986 tape of mail collaborations, *Piñata Party*, in a plastic fruit container filled with stickers, toys, and candy. In perhaps the simplest concept of all, Eugene Chadbourne's 1991 set *Dirty Sock* comprised four tapes literally stuffed into a dirty sock.

The mail also gave artists opportunities to work together from afar. Home tapers swapped sounds and ideas, overdubbed tapes back and forth, and made chain-letter compilations, with each recipient adding music before forwarding a cassette. Often, these artists never met face to face, adding a level of mystery to these message-in-a-bottle missives. Who is this person? How did they do this? Such questions might never get answered even as close relationships were formed. Many home tapers also saw the technical limitations of the cassette tape as an opportunity. Its lower fidelity helped them create textured, abstract experimental music, noise, and other unclassifiable forms of sound art. The fragility of the format was a tool for anyone who liked to play with concepts of time, and the fact that you could record over existing sounds added ephemerality. "It's scratch paper, sketches, temporary, which is a legitimate art form in itself," says James. "You can reuse it. With other formats, there's no second chance."

OP and Castanets continued to cover cassettes until 1984, when the magazine closed and spawned two similar publications, *OPtion* and *Sound Choice*. But the ripple effect of their tape coverage persisted. Hal McGee was already making music with his partner Debbie Jaffe in Indianapolis, Indiana, when he discovered *OP*. The couple would buy a bag of twenty-five blank tapes for five dollars from a nearby drugstore and record with a shoebox-sized deck meant for documenting lectures and meetings. *OP* inspired them to bring their music to the outside world, connecting with every other tape artist they could find. "I actually took it seriously, that independence of doing it on your own and being true to your artistic vision," says McGee. "So I thought, 'Okay, let's create a kind of community and highlight what's important.'"

McGee and Jaffe started a label and distributor called Cause and Effect, sending out mail-order catalogs so people could buy tapes by multiple artists instead of having to contact each one individually. At first Cause and Effect bought handfuls of each title, but eventually they also requested single master tapes from artists, which they duplicated themselves, making as many as needed to satisfy orders. "From about 1986 to 1988, we sold over 5,000 cassettes that we dubbed ourselves," remembers McGee. "Our decks were constantly going, and after each dub we'd pop the little tab off the top so you couldn't accidentally record over the tape. People would come to visit and see our floor just covered with those things."

McGee has made his own sound art too, some released under the name Dog as Master, and collaborated with scores of other home tapers. His work is driven by the ideas that anything in life can be art—he often carries his tape recorder with him to work to record mundane events for source material—and that personal expression shouldn't go unheard just because the mainstream ignores it. "One of my great joys is to try to make my art and my life as close together as possible," McGee says. "When you hear one of my tapes, I want you to right away go, 'Oh yeah, that sounds like him.' That's the difference with cassette artists. What sets us apart is individual personalities, our viewpoints, our experiences."

Individual points of view abounded in underground cassette culture, where the range of creative personalities seemed infinite. In Torrance, California, a reclusive artist named Stanley Keith Bowsza called himself Minóy and made tapes while rarely leaving his house. He sometimes composed by playing cassettes on multiple boom boxes, then recording that sound onto a Walkman, controlling the mix by manipulating the volume levels on his gear. He built a discography of over 100 releases, including fifty made in just 1991 alone. After Bowsza died in 2010, an artist he often worked with, Phillip Klingler, was bequeathed over 300 tapes of his material, which he is still patiently releasing one by one.

Hours north in San Jose, Don Campau worked as a supermarket clerk and spent his downtime writing songs and releasing tapes with names such as *Mr. Full Time Vegetable* on his own Lonely Whistle label. His tireless networking and collaborations made him one of the most active figures in the scene. He would even give tapes out at his day job to people unfamiliar with experimental music and bring copies to his local library. In Los Angeles, Dino DiMuro crafted compositions with guitar, keyboards, and answering machine messages during breaks from his film industry gigs. In Eugene, Oregon, Heather Perkins spent time between jobs mixing home-recorded songs with brutally honest mono-logues. In Boulder, Colorado, the ensemble Walls of Genius released over thirty tapes of pop and classical deconstructions while performing under a series of different names, as if they were a scene unto them-selves. In New York and Seattle, songwriter and multi-instrumentalist

Sue Ann Harkey made avant-folk on her label Cityzens for Non-Linear Futures and improvised with her group Audio Letter, producing collaborative tapes with artists across the underground.

In Des Moines, Iowa, Zan Hoffman discovered *OP* in the early 1980s, and quickly began making wildly varied music with tapes. He was especially devoted to working with every tape artist he could find. "I thought: all my home taping friends have unfinished recordings lying around," Hoffman told Campau in 2011. "In order to solve my anticipated issue of limited audio sources, I could mine the most interesting international source I knew of: my home taping pals!" For his 1985 release *40 x 40*, he took forty-second snippets from those pals and collaged them into new works. Along with his solo music under the names Zanstones, Zanoisect, and Bodycocktail, Hoffman churned out tapes so quickly that he released 120 in 1988 alone, eventually amassing over 1,000 on his label ZH27. Klingler estimates that his own source material has been used on at least fifty of Hoffman's releases.

"Zan Hoffman is probably the man who defines home taping in all its aspects," fellow artist Tom Furgas told Campau. "His energy and enthusiasm are enthralling, inspiring, and nearly superhuman." "How I've developed, my varied emotional states, my evolving network of collaborators, the march of technology and equipment through my life—all of these are documented in their own way across over 1,000 releases," said Hoffman. "I am proud how [my] releases are getting at providing the answers to questions nobody else is asking."

In New York City, artist Gen Ken Montgomery watched tapes from all over pile up in his apartment and got frustrated by the challenges of getting them into stores. So in 1989 he started his own shop, Generator, to, as he wrote in Robin James's *Cassette Mythos*, "see if it was possible to get new audiences interested in the worldwide cassette underground." Manning the counter in off-hours from his photo-lab day job, Montgomery presented Generator as a cassette gallery, hanging tapes on the wall with Velcro, spreading players and recorders around, and positioning speakers everywhere, including the bathroom. He also posted signs explaining that the store was a place for discovery, so don't expect to recognize or even understand all the music inside. Generator struggled to break even and lasted for only a year, but, as Montgomery wrote later, "I think the greatest achievement of Generator was exposing newcomers to this audio revolution, planting seeds to help the network grow."

Growing up in the Washington, DC, area, Leslie Singer spent her youth in punk rock bands, but by the mid-1980s she had gravitated toward noisier extremes, and cassettes freed her to follow that muse. "Some part of me thought, 'Melody makes it too easy,'" she remembers. "I like to have something to push up against, something that has a little grit in it." Under the name Girls on Fire, Singer made tapes with boatloads of grit, with titles such as *Diary of a Shit Eater* and *Confessions of a Shit Addict*. "It all sounds like it's going to erupt from the tape deck and shred speaker cones and eardrums," wrote Hal McGee about her music. Cassettes gave Singer the chance to disseminate material that commercial labels would never touch. "I knew that people wouldn't put this on and just go about their business," she says. "I think that what I do demands something from the listener. I try to make it funny and interesting, but I'm demanding something."

Driven by a similar spirit, Amy Denio began making her own tapes in Seattle when she discovered the cassette underground. "I've always been suspicious of the music industry, and this way I could completely maintain the means of production," she explains. "Then it turned out to be this really cool social network, which I had no idea existed." Her first tape, 1986's *No Bones*, offered a dizzying hour of avant-garde jazz, playful theatrics, and abstract sound poems. The cover included a photo of Denio in the hospital, after a bike accident that left her with a bandaged head but somehow no broken bones. Self-released, *No Bones* soon circulated widely thanks to Sound of Pig, a label and distributor run by Al Margolis in New York. "Al got my music out to the most obscure corners of the world," she remembers. "Like, suddenly I was getting mail from Israel, and I thought, 'How did my music get to Tel Aviv? That's insane.'"

Such a story was routine for Margolis, who had his ear on seemingly everything in the cassette underground. In 1984, he started Sound of Pig and his own music project, If, Bwana (an acronym for It's Funny, but We Are Not Amused). "I'd been buying cassettes from people for a couple of years," he told author Thomas Bey William Bailey in 2012. "And it just dawned on me that if I started a label, then I'd have something to trade instead of spending money." Margolis spent most days dubbing tapes at home or on decks he carted to his job. In five years, he released 300 titles, and in 1985 alone he made 3,000 copies of tapes. When a musician didn't provide cover art, Margolis took up that task too, cutting and pasting images from books and magazines and duplicating them on his own home photocopier. "I'm not trying to make money," he told the podcast *Noisextra* in 2022. "This is trying to contact people. If you just had one person you traded a tape with, that's it. You made a connection with another person."

Many other cassette labels thrived as well. In Memphis, Tennessee, Chris Phinney released compilations of experimental music from Germany, Spain, Czechoslovakia, and Japan on his Harsh Reality imprint. In Los Angeles, Randy Grief's Swinging Axe produced scores of international noise cassettes. And in Massachusetts, Ron Lessard sold tapes through the mail—mostly by advertising in the collectors' magazine *Goldmine*—before opening his own store and label, RRRecords. For one series, *Recycled Music*, Lessard invited artists to record onto discarded rock and pop tapes; sometimes the original music bled through their noise.

Around the same time, Lessard created *RRRadio*, soliciting tape material for live audio collages. Similar radio shows around the world espoused the gospel of home tape artists. GX Jupiter-Larsen ran one in Vancouver called *New Sounds Gallery*. David Lichtenverg, also known as Little Fyodor from Walls of Genius, hosted *Under the Floorboards* in Denver. In Allentown, Pennsylvania, Joe Schmidt's *Home Taper* show solicited cassettes, played them, then made copies to exchange for future submissions. In Cupertino, California, Don Campau hosted *No Pigeonholes* on KKUP for decades. "I give everyone airplay no matter the style, no matter if they suck," he said in Jerry Kranitz's book *Cassette Culture*. "That being said, I am always amazed at the amount of talent and creativity out there."

The cassette also proved to be a ripe format for audio magazines. In 1973, conceptual artist William Furlong began the series *Audio Arts*, in which he interviewed other artists, then mailed the tapes to friends and eventually subscribers; it lasted well into the 2000s. "It became apparent to us," he told the *New York Times*, "that none of our interests were being met by any traditional arts publications." In New York, the cassette periodical *Tellus* mixed all kinds of music and audio art in bimonthly editions for a decade. In Boulder, Colorado, Joel Haertling published the compilation-oriented magazine *Zamizdat Trade Journal*. And Robin James ran the tape series *Audio Archive*, asking a set of standard questions to many home tapers and publishing the answers on cassette.

It seemed everyone in the 1980s cassette underground was busy in multiple ways, another testament to the strength of community in the scene. "There were no passive spectators," insists Jupitter-Larsen. "You were an artist, a performer, you had a radio show, you had a label, you wrote for a zine, you published a zine, you were a promoter. Everyone participates, everyone's equal, no heroes, all that." "There was a lot of consciousness of other artists," says Tony Coulter, a tape collector who plays underground tapes on his show on WFMU in East Orange, New Jersey. "People were playing to a very informed crowd rather than some hypothetical mass audience."

The 1980s cassette underground stretched beyond America. In Japan, acts such as Solmania, Masonna, and MSBR often recorded homemade noise on cassette, spreading it through Japan and around the world via mail. In 1983, Hijokaidan released a ten-cassette box set of early work called *Gokuaku No Kyoten*; Merzbow, the intensely prolific project of Masami Akita, began in the early 1980s by making what he called "cheap cassettes which could also be fetish objects"; and Marble Sheep and the Run-Down Sun's Children put out twenty-one live tapes in the late 1980s, most released before the band had even recorded a studio album. Packaging was vital too: take the Gerogerigegege's *This Is a Shaking Box Music (You Are Noise Maker)*, a metal box containing 100 empty cassette shells alongside photographs and "trash." Due to the absence of any music, the listener was supposed to shake the box around to make noise. Three copies were made.

In the Netherlands, Frans de Waard discovered cassettes as a teenager when reading about tape labels in the music magazine *Vinyl*. "It was a like a light went on in my head," he remembers. "All of the implications that came with the whole world of cassettes was immediately clear to me." De Waard started making experimental music with tapes, learning to change the speeds of his sounds by pressing play and fast-forward at the same time. His work got a boost when Dutch radio host Willem de Ridder played one of de Waard's tapes on the air. "It sounded like a bunch of hiss over the radio," de Waard said. "Nevertheless, hearing my music on the airwaves spurred me to create my next composition."

It also spurred de Waard to chase every bit of information he could about tapes, writing to or telephoning any name he came across—"much to my parents' misery," he remembers. In 1983, at just eighteen years old, he published *De Nederlandse cassette catalogus* ("A catalog of cassette releases in the Netherlands"). "I was suddenly being interviewed on the radio and for magazines," he wrote. "I had blossomed into an authority on the subject—at least in general perception." He would publish three more editions of the catalog, while also starting his music project Kapotte Muziek and his tape label Korm Plastics. For de Waard, affordability makes tapes their own kind of medium. "If you had to fork out thirty dollars per record, you might give up after five records that sound the same," he explains. "But if a cassette costs you five dollars, you can follow the progress of a musician over time. If I could afford to release all my music on vinyl, I would still not do it. Cassettes are a way to show that this is what I do. This is my development."

Also in the Netherlands, poet and mail-art enthusiast Rod Summers was drawn to tape in the 1960s. While serving at an army base in the city of Maastricht, he acquired a tape recorder, a tone generator, and an editing block and got to work. "I make audio art and not music," he told Thomas Bey William Bailey in 2012, "because I'm an artist and not a musician." In the 1970s, he created the VEC (i.e., Visual, Experimental, and Concrete) Audio Exchange to, as he put it, "inform artists working with the medium of audio about the contemporary situation." Each VEC cassette contained twenty contributions and was only available via trade for more source material. Over five years and sixteen releases, Summers's project attracted 180 artists from twenty different countries. "The global aspect of the VEC was, in part, a means of achieving an understanding of creative humanity," he insisted.

In Italy, Vittore Baroni also found mail art in the 1970s, publishing a magazine called *Arte Postale!*, which led to him correspond with home tapers. "Receiving those early homemade cassettes was an illuminating experience," he told Don Campau in 2011. "I realized that I also could do something like that, even if I was tone deaf and I never learned to play a guitar." Under the name Lieutenant Murnau, Baroni used tape decks and reel-to-reel players to mix field recordings, instrumental sounds, and vinyl samples. In 1981, he cofounded TRAX, hoping to mine the best aspects of mail art and the cassette underground. He considered the label's releases to be more than just audio; tapes often came with postcards and small pieces of art. He also considered TRAX a parody of modern industry. "Our products could often be fragmented and reassembled in different combinations," he told Thomas Bey William Bailey. "This was intended to . . . include some form of participation from the part of the buyer, but it also reminded you of the factory assembly line." His "modular system" dictated that anyone who organized a TRAX release was a "Central Unit," collaborating with contributors who were considered "Peripheral Units," in hopes that his idea would be adopted by thousands of artists across the globe. Baroni might not have achieved that goal, but by the time TRAX ended in 1987, over 500 people had participated in nearly 100 different projects.

In Germany, artist Graf Haufen (a.k.a. Karsten Rodemann) ran Graf Haufen Tapes, at first releasing his own minimalist synth music, then welcoming other artists and shifting more toward industrial noise. He also published a magazine called *Die Katastrophe* dedicated solely to cassettes. In Belgium, musician Alain Neffe produced compilations of international artists on his Insane Music label, in series called *Insane Music for Insane People* and *Home-made Music for Home-made People*. Neffe accepted any music regardless of quality. "If the buyer didn't like the music, he or she could wipe it out and record something else on it," he explained to Campau. In Australia, Bruce Milne and Andrew Maine created *Fast Forward*, a magazine devoted to cassettes that was eventually published on cassette (later inspiring Bruce Pavitt to do the same in Seattle with *Subterranean Pop*).

In the UK, cassette culture was partially driven by print. In 1980, writer Mick Sinclair started a tape review column in the music weekly *Sounds* called Cassette Pets, while at least one independent fanzine, *Stick It in Your Ear*, covered only tape releases. Barry Lamb's Falling A label distributed tapes and eventually became a store, while Andi Xport started a series of compilations called *International Sound Communication*, highlighting music from, he claimed, almost every country on the globe. "Absolutely nothing was rejected," he told Campau. "I stopped because I just could not keep up with all the mailing and was spending four to five hours a night writing letters, doing post and duplicating tapes."

Though based in California and Oregon, Archie Patterson often focused on tapes from outside the United States. He took his early 1970s radio show, *Eurock*, and turned it into a magazine that reviewed cassettes, then a distributor. He started with a tape by the Plastic People of the Universe—smuggled from their locked-down country of Czechoslovakia—and later helped boost a burgeoning electronic-music tape scene in Mexico. "There's a certain quality about [this music], the way it was composed and sounded, which could have only happened with the advent of cassette tapes," Patterson wrote in the notes to a Eurock compilation. "The cassette format facilitated the imagination and freed up musicians to do their own thing, their own way, and be heard."

It would be impossible for a single guide to cover all these overlapping underground scenes, but Robin James gave it a try. In the early 1990s, he expanded his *Audio Archive* zine into a full-fledged book called *Cassette Mythos*. With contributions from cassette makers, traders, and proselytizers, it's part resource and part manifesto, a far-reaching collection of statements from artists whose lives were changed by tapes. "The cassette is the counterculture's most dangerous and subversive weapon," Hal McGee wrote in the book's intro. "It is a threat, an incendiary device. . . . What makes the worldwide cassette movement so unique is that it is a society of real participants, producers instead of passive consumers."

Some of these prolific home tapers went beyond using the cassette to release music and transformed it into an instrument in its own right. In the mid-1980s, Robin James staged concerts where he turned a stack of cassettes into a makeshift band. He asked attendees to participate by bringing their own tape players. "They were spread out all around and the lights were down low," he remembers. "Then we passed out the tapes and everyone pushed play at the same time." James's experiment echoed one of the earliest uses of tapes as instruments: Filipino composer José Maceda's *Cassettes 100*, comprising 100 tapes that Maceda had filled with instrumental sounds and field recordings. During the piece's 1971 debut at the Cultural Center of the Philippines in Manila, 100 participants walked around with Maceda's tapes blasting from handheld players, creating an ever-changing three-dimensional audio experience.

A crucial pioneer in the use of cassettes as instruments was German composer Conrad Schnitzler. A member of Tangerine Dream and Kluster—bands who helped define a repetitive psychedelic sound known as Krautrock—Schnitzler was fascinated by cassettes. At one point, he walked around Berlin strapped with a string of tape players wired to a speaker on his helmet, composing on the fly. Soon after, he sought a way to perform his music without having to play everything himself, hire multiple musicians, or haul stacks of gear. His solution was a new instrument he called the *Kassettenorgel* (cassette organ), comprising two cabinets holding six tape decks wired together. In concert, Schnitzler played tapes filled with sounds from his synthesizers, performing alongside them. Hoping to replicate this performance around the world without traveling himself, he created the Cassette Concert, wherein others could play his prerecorded cassettes on multiple decks. Each tape held a portion of the composition—roughly equivalent to a line of notation in a traditional score—so that playing all of them in sequence represented a performance of the full piece.

Allowing others to perform his work introduced subjective interpretation and chance improvisation, which Schnitzler valued. Different performers, or "conductors," could vary the piece based on where they placed the decks, how loud each one was, and when each was tape played. They could also introduce new cassettes into the mix and even allow audience participation. "The Cassette Concert transforms a static recording into an evolving intermedial event," wrote Gen Ken Montgomery, Schnitzler's first authorized conductor, in *Cassette Mythos*. "Creativity is passed through the medium to encourage and inspire the inherent creativity of the audience."

About a decade later, an American rock band had a similar idea. As a teen in the late 1970s, Wayne Coyne was cruising a parking lot before a Kiss concert and became entranced by a swirl of sound from 8-track tapes booming out of car windows. About twenty years later, a decade into the career of his group the Flaming Lips, he and bandmate Steven Drozd imagined gathering a bunch of cars and giving each driver a tape containing a segment of a multipart song. The pair made upward of forty such tapes, working meticulously to cut and paste music onto each. "I went out and bought fifty or sixty little tape decks so I could do a dress rehearsal right in my house," Coyne told author Mark Richardson.

In the spring of 1996, cars packed into a mall parking garage in Coyne's hometown of Oklahoma City for the first of what the Flaming Lips would call their Parking Lot Experiments. Once Coyne yelled "Play!" each tape announced a number in sequence, then a round of music erupted in the garage. Drivers were encouraged to get out and walk around, experiencing their own personal mix of this collective composition. "Some of the tapes don't have much on them at all, but as a big piece they all work together," Coyne recalled to Richardson. "Sometimes driving around, we'd pop one in and wonder 'What's on that thing?' And it would announce, 'This is tape #28,' and for ten minutes there would be absolutely nothing, and then there'd be some scream or something."

After staging a few of these events, the Flaming Lips reworked their idea into a more portable form. Their Boom Box Experiments featured forty tapes loaded into forty boom boxes. Audience volunteers pressed play simultaneously, and Coyne and Drozd conducted the crowd, waving their hands up and down to direct changes in volume. One attendee described the results as "a cyclone of police sirens, dogs barking and holy 'ah!'s." As with Schnitzler's Cassette Concerts, no two performances were alike, with timing, volume, spacing, and the anomalies of each tape and player making the results unpredictable. "What people like about it," Coyne told *SF Gate*, "is that it might screw up."

Schnitzler's and Coyne's use of tapes as instruments involved a lot of spectacle. Scores of other artists have included cassettes in their arsenal in ways that are perhaps less showy but just as innovative. Growing up in the late 1980s in Michigan, Aaron Dilloway was watching a video of a Throbbing Gristle concert when he noticed them playing cassettes on stage. Having made cassettes of his band Galen and sold them at local shows, he decided to incorporate tapes into his own work. At first, he made loops on 8-track tapes, inspired by Mick Jagger's cycling score for Kenneth Anger's 1969 experimental film *Invocation of My Demon Brother*.

Eventually, Dilloway used cassette tapes in his work as well, aided by the acquisition of a C1, an early 1980s tape deck created by the Library of Congress for use by the blind. "That was just life changing," Dilloway remembers. "You can vary the speed, you can flip sides immediately, you can change the tone of the tape. I still think of that as the best musical instrument ever." Gradually adding other decks, he came to see them as instruments with different capabilities, like models of guitars. "There are ones where you can push the play button down and hit fast-forward or rewind and hear the song being sped up or down," he explains. "There are ones where when you hit stop, it'll stop on a dime. There are ones where you can push the stop button down in a way that slows it down a little bit first before it's totally stopped. Every single one of these decks is different."

Dilloway uses both cassettes of sounds he makes and prerecorded tapes found in thrift stores and other random places. For years, one of his main sources has been a box of answering machine cassettes, which usually start with five seconds of silent, unrecordable leader. "I originally found that annoying," he says. "But I realized that if I started using them with enough cassette players at once, that those spaces kind of create a rhythm of their own." His collection of tapes helps him compose on the fly, accessing sounds he knows from years of use as well as grabbing random tapes and letting chance dictate results. In that regard, his found cassettes are particularly useful. One such tape, discovered at a secondhand shop and featuring the music of a church choir, remains in

Aaron Dilloway, who uses cassettes
and decks as instruments, works with
magnetic tape during a residency at
New York's Wave Farm.
(Photo by Bryan Zimmerman)

his repertoire decades later. "It's just the spookiest tape," he enthuses. "I
still don't really know what's going to come out when I use it. I pulled it
up at the end of a set once and it was a version of 'Amazing Grace,' and
I thought, 'That's maybe a little too heavy to be pulling out right now.'"

Of course, Dilloway could do all this in a simpler, less cumbersome
manner with computer software. "I can't work like that," he counters.
"I'm not as inspired sitting in front of a computer, and I don't have any
desire to be working with numbers and codes and stuff like that. I need
to be able to touch it with my hands and work with the sounds that way.
Half the time, I have my eyes closed when I'm playing live, so I need to
have knobs and be able to feel what I'm doing."

Howard Stelzer, who has made music with cassettes since the early
1990s, feels similarly. "At this point, it's how I think about expressing
sound," he says. "It's so bodily for me to press the pause button and lean
on the reels of a tape. It might be like the way a singer thinks of his
voice." Stelzer's fascination with cassettes began when he grew up in
Florida, ordering noise tapes based on descriptions in the RRRecords
catalog and ads in *OPtion*. In high school, his band spent afternoons
banging on junkyard debris in his family's garage. Recording their din

Howard Stelzer performs with his arsenal of cassette tapes and decks at Washington Street Arts Center in Somerville, Massachusetts, 2015. (Photo by Howard Stelzer)

on a handheld cassette recorder thrilled him. "The idea that we could make sound, then immediately play it on a stereo was a revelation," he remembers. "And what the recorder did to change the sounds was a compelling mystery to me."

Stelzer began recording all kinds of sounds to cassettes, launching a lifelong practice of carrying recorders with him everywhere. His friend Robert Price of experimental group Kreamy 'Lectric Santa suggested Stelzer make collages of these tapes using a dual-cassette deck, sparking another revelation. "I couldn't really control where the edits were; I was just throwing in tapes as fast as I could and hitting pause as fast as I could," he remembers. "What came out blew my mind." Moving to Boston in the late 1990s, Stelzer found a community of improvising experimental musicians to collaborate with. At first, he brought tapes to shows in a trash bag, deciding on the fly when to play each. But now he writes his music beforehand, putting specific sounds onto cassettes and playing them in a predetermined order. Sometimes he'll put a version of the same sound on eight tapes, then play those on decks positioned at varied distances, creating a kind of cassette symphony.

"I still love the unpredictability of tapes," he explains. "I enjoy getting a sound and playing it out of one Walkman and recording it onto another, and then taking that tape and playing it outside as I'm walking my dog and recording that, and then taking that tape and playing it in my car and recording that. As the sound is changed by the environments and tapes and decks, I can let the music make itself, and I end up with something that I couldn't have predicted at the beginning."

The unpredictability of cassette tapes fascinated Michael Anderson when he was growing up in the 1980s in Massachusetts. At age twelve, he got his first deck, using it to interview fellow classmates, then manipulating the sound by opening cassette shells and mangling the tape inside. A few years later, he realized he could make his own loops by taking cassettes apart and inserting small lengths of tape. "I'm still using some of the tape loops I made back then," he says. "I love how they change over time from use, in the quality of the tape and the tone. And the tape starts to fall off the reel a bit and creates gaps in the loops, which changes them too."

In the mid-1990s, after moving to Bloomington, Indiana, Anderson adopted the name Drekka (Icelandic for "drinking") and developed a heavily physical performance style. He would lean forcefully onto his table of tape players—including the same Library of Congress decks that Aaron Dilloway uses—and stalk around stage. "It's very ritual, based around the placement of the objects and how they move around the table and the stage," he says. "People ask why I carry around all this gear when I could just use a laptop, but it's important to my performance to have these physical objects to interact with." In one of his bands, Racebannon, he would sit on stage and tape the music, then play it back out while editing it on the fly. "I'm not very academic—it's pretty intuitive for me," he admits. "Because of the physicality and mutability of tapes, they're great for my kind of intuitive approach."

The intuitive approach of Norwegian musician Sindre Bjerga began during childhood, when he saved cassettes his family made. "I still have this tape from 1979 where I am practicing my reading skills by reading out loud the blurbs in some comic books," Bjerga recalls. "And my dad is playing Bach pieces in the background." Bjerga first worked artistically with cassettes in the mid-1990s, using a recorder and tapes his mom used when she worked as a secretary. Along with found sounds, he records his own instruments onto degraded tapes to add texture to his drones. In performance, Bjerga employs multiple decks, microphones, and some children's toys to make improvised collages. "I use crappy mics in different positions to filter the sounds from the tape players, in addition to physically messing with the tapes while playing," he explains, "such as 'scratching' with the tapes in one of my Walkman players." Once, he wrapped a player in tinfoil, and discovered that "it sounds like there's a thousand angry wasps inside it!"

Bjerga considers his use of old, found tapes a form of recycling. One particularly durable example is a tape he found at a flea market that includes sounds of a woman singing karaoke and people playing a board game. From the writing on the case, Bjerga estimates the tape is over half a century old; he has used it in at least fifty performances. "You could say there are 'enough' recorded sounds in the world," he insists. "So why not use sounds that are already recorded?"

Similar philosophical questions arise in the work of Jason Zeh. The Ohio-born, Kansas-based artist explores the question of, as he puts it, "What does cassette technology want to say, and how can I help it speak for itself?" Growing up, Zeh watched his father dub records he checked out from the library, using a photocopier at work to make cassette covers that matched the LPs. Soon, Zeh began making art for his own imaginary albums, later recording simple songs to cassette. In high school, he took a microcassette recorder, put it inside a banjo, and captured sounds from the body of the instrument, launching himself into a lifetime of tape experimentation.

In college, Zeh sometimes composed by laying different cassettes out in front of him, creating a physical score with instructions for transitions scrawled on the shells in marker. Over time, he prioritized the medium as an object rather than an audio container. "I was dogmatically opposed to using prerecorded sounds," he explains. "I wanted to push myself to use tapes in a way that was as opposite of their intended use as possible. My thinking was that tapes are a communication technology intended to convey some message. When that message is the focus, people fail to hear the essential qualities of tape. I wanted to avoid the meaning, to focus on the noise that would otherwise go ignored."

To achieve this goal, Zeh continually devises new ways to use cassettes. He cuts holes in them so he can loop the tape through multiple shells. He holds tape players over open flames, producing a high-pitched sound from the melting plastic. He scratches blank tapes, adds magnets, and even removes the tape entirely, replacing it with contact microphones and pieces of plastic, metal, and sandpaper. "I was hoping to discover something essential and true about [tape] materials," he says. "To make those materials speak for themselves."

In the age of pristine digital audio, when it's easy to ignore the mediums that deliver music, tape is even more valuable to Zeh. "If technology can fade into the background of our lives, then it is easier to ignore the growing mound of e-waste that is a necessary consequence of technological advancement and planned obsolescence," he posits. "Drawing attention to and repurposing the technological trash around me is important to engaging with that reality. The only way I could think to do that was to bring to the forefront the noise produced by tape technology."

For some artists who use cassettes, the format offers a chance to be not just philosophical but personal. While living in London in his twenties, Japanese sound artist Aki Onda wanted a camera but couldn't afford one, so he bought a tape recorder instead. "I think that affected the way I captured my field recordings, using a tape recorder as if it were a camera," he recalls. "I recorded any sounds around me whenever they caught my attention, and I consider those memories of my personal life." At first this was more a therapeutic process than an art practice, but gradually Onda saw tapes as creative objects, making collages on their cases with drawings, memos, and scraps cut from magazines.

In the early 2000s, Onda began composing with his cassette recordings, launching a series of releases he calls *Cassette Memories.* He was inspired by turntable artists, experimental filmmakers, and hip-hop musicians. "I was especially interested in how they deal with their own memories," he says. "They quote from the Black cultural heritage and beyond, take the sounds away from the original context, and create new compositions. . . . They archive past memories and edit, rewrite, recite, and remix them for a future use." Eventually Onda started performing *Cassette Memories* but realized the project didn't fit in conventional concert settings. Instead, he sought out places that had history

79

A selection of hand-decorated cassettes and Walkmen used by Aki Onda, some of which he uses in his *Cassette Memories* project. (Photo by Aki Onda, courtesy of the artist)

behind them—as he puts it, "a space that has its own memories . . . a historic building, abandoned factory, old theater, even a street corner. It's a strange ritual. . . . The result is invisible, but it is as if live memories awaken sleeping ones." Such locales include a 2012 performance at Cour Carrée (an outdoor section of the Louvre Museum) and a 2018 appearance at Poland's Festiwal Sanatorium Dźwięku, held inside a nineteenth-century hospital once used to treat tuberculosis. The history of this kind of cassette-based performance fascinates Onda as well: in 2019, he directed a performance of José Maceda's early-1970s piece *Cassettes 100* at the Kanagawa Arts Theater in Japan.

In working with cassettes, Onda hopes to take personal moments and divorce them from their original context, much the way memory manipulates experiences. The distortion and deterioration of cassette tapes make them a perfect tool. "The sound can be narrowly compressed and flattened frequency-wise; the texture can be rough, coarse, and grainy," he says. "It's not a medium to re-create the aural reality as it is. In a sense, I exploit or maximize that nature."

Maximizing tapes is integral to the work of UK-based artist Stuart Chalmers, a self-described "sound scavenger." He cracks them open to create short loops; takes out the tape and crushes it with gravel; captures improvisations and field recordings on cassettes; pulls sounds from tapes he finds in trash bins; and even records the noises of tape player mechanisms themselves. "I like to keep using [tape decks] until they completely stop working," he says. "When they start failing or not doing what you expect is when things open up and get interesting, with chance and chaos being a positive factor in shaping the recording." In his *Loop Phantasy* series, he blended tape loops of different durations, focusing on the ways they drift in and out of sync. For his retrospective work *A Life in Sound: Diaries 2001–2020*, he collaged cassette recordings from over two decades, including personal diaries, travelogues, and the voices of friends and family. "There may seem to be more limitations using [cassettes]," Chalmers says. "But it means you have to be more creative and work with the restrictions."

Also in the UK, Joe Murray makes similar experiments with cassette tapes—but the kind he prefers are smaller. Most of his work, released under the name Posset, is done with microcassettes in a Dictaphone, a device designed for voice recording. "With Dictaphones you have huge amounts of control," he says. "It's right down to the amount of pressure you put on a button. You can nudge a tiny blip of sound forward or backward, dragging it out over the tape head in such an intuitive and human way." Murray spent most of his early years in rock bands, but he was always interested in cassette tapes. A breakthrough came when he heard "Poledo," the tape collage Lou Barlow included on Dinosaur Jr.'s album *You're Living All Over Me*. "I fell in love with the dusty roar of the tape," he remembers. "The way it descended into slow motion, the fucked-up voices and heavy, hissy tape loops. It was magical and mysterious but strangely attainable too."

In concert, Murray usually uses only tapes, with no effects or other sound sources. His music mixes audiobooks, street recordings, and found sounds—often hilarious ones, such as a tape from a company's customer service department filled with phoned-in complaints. His short sets often start with an explanation of how he does what he does, to make his process less mysterious. "I want what I do to be accessible and inclusive," he insists. "In many ways my music is pretty meaningless. I'm just reveling in glorious sound."

Accessibility also marks the work of Randall Taylor, who makes cassette-based soundscapes under the name Amulets. Growing up in the 1990s, he taught himself to tape songs off the radio. As an adult, turned off by computer software that took too long to learn, he returned to tapes, teaching himself to make loops with the help of instructional videos on YouTube. "Tapes are just so easy to use and accessible, and I already loved going to thrift stores," he says. "So I figured, now I'm just looking for something else there." His treks are often fruitful: many of his songs extract sounds from old books on tape. "I look around in my room sometimes and think, 'What did I get myself into?'" he says, pointing to his shelves full of decks, tapes, and cases. "I was really into Legos as a child, and I have this weird adult's musical Lego set now. I just like to build things, connect it all, and make something new."

On his YouTube channel, Taylor posts videos demonstrating his various tape techniques. In 2019, he streamed a live video of a "self-destructive tape loop," which played on a Library of Congress deck for an hour until it collapsed. Positive response to such clips has led him to give tape-loop lessons over the internet via Skype. "I think the simplicity really piques interest," he says. "This weird little physical five-second loop can really change a lot of people's perspectives." Taylor keeps only one copy of each loop he uses for a composition, giving his pieces a literal shelf life. "I've had songs where the tapes have worn themselves to a point where they sound very different," he explains. "I even have songs that have decommissioned themselves from my set because they just wear away, and I think, 'Well, I should've stopped playing that song anyway.' But what told me is the tape itself. It basically said, 'You're done playing this song.'"

The 1980s cassette underground that included or inspired so many of the above tape artists started to slow a bit as the decade ended. Rising rates at the post office, gluts of tapes that were hard to sell, and general burnout all took their toll. "There seemed to be a lull in the late '80s," Al Margolis told *Collectors Weekly* in 2015. "I don't know if that was due to CDs coming in or just the flagging energy of the home tapers. A lot of the zines dried up. I got tired, and at one point, I just stopped." "I started working for Staalplaat [in Amsterdam] in 1992, and behind the counter we had this wall of cassettes," remembers Frans de Waard. "No one bought them, so we took them down and put them in a bin, selling for one guilder [a Dutch coin worth about fifty cents] each. Charles from

Soleilmoon Records did the same thing. He had bins full of Throbbing Gristle cassettes that he couldn't sell, so he finally said, 'Take whatever you want.'" "What happened was you lost your first major generation of people you traded with," Zan Hoffman told Jerry Kranitz. "And most people weren't willing to find an entirely new generation of contacts. But there's still interesting people. There always will be. You just had to take effort to weed through another whole range of artists."

Indeed, new underground scenes emerged in ensuing decades. Participants extended into uncharted musical areas, but they still hand-dubbed tapes, made their own cover art, and traded and collaborated across states and continents. In Japan, Nakajima Akifumi, who made music under the name Aube, ran a label called G.R.O.S.S. through which he sent noise tapes to North America and Europe, often in trade for the same from artists in those countries. The wildly creative presentation of cassettes in Japan also persisted. G.R.O.S.S. released Masonna's *Passion of the Rubbers* in a rubber case with belts tied around it; Masonna himself released a compilation, *Gomi-Atka*, that came taped to a discarded wad of tinfoil; and MSBR put out a tape called *Fracture of Silence* housed in a huge piece of foam that resembled a burnt loaf of bread.

In America, a scene coalesced in the Midwest via labels such as Hanson (run by Aaron Dilloway), Freedom From, Chondritic Sound, and Gods of Tundra. But other imprints around the country (Shrimper, Union Pole, Hospital Productions) and the UK (Chocolate Monk, Betley Welcomes Careful Drivers, Fourth Dimension) also made tapes that pushed the limits of musical convention and volume, packaged in art as visually arresting as it was boundary pushing. Perhaps the most wildly prolific of these was American Tapes, run by musician John Olson in Michigan. He launched the imprint in 1995 while working at the photocopy chain Kinko's, which gave him access to leftover materials and free photocopies. By the end of the decade, he had released 100 titles, and by the mid-2010s that number had jumped to 1,000. Most of his releases had distinctive hand-collaged or silk-screened covers, often stamped with a date and a catalog number. The roster of artists was massive, including many pseudonyms for Olson himself. "Some cats just use their name for a jam; I personally don't like doing that so much," he said in 2020. "It's music. It should be its own world. It should be a gateway into escapism."

Like many of the experimental tape labels of the 1990s and 2000s, American Tapes also put out LPs and recordable CD-Rs. For many underground artists, the latter format served a purpose similar to that of the cassette tape in previous decades. CD-Rs were even easier to make at home than tapes, since computers did most of the work for you, and were considered just as valid a vehicle for cheap music dissemination. "When you just are a tape label, it's limiting," Olson insisted. "We wanted it to always be the same style throughout the formats . . . you shouldn't think that it's a whole different thing." Still, tapes proved to be a perfect vehicle for anyone who wanted to carry on the tradition of the 1980s underground. "I can't imagine ever fully stopping tapes; they are the symbol of the underground," said Dominick Fernow of the noise label Hospital Productions. "What they represent in terms of availability also ties back into that original noise ideology. Tapes are precious and sacred items."

And even if the original generation of underground tapers dissipated, many pioneers stuck to it. After shifting from tapes to CD-Rs in the 1990s, Frans de Waard became entranced again by cassettes in the late 2000s; one of his current projects, Modelbau, uses tape loops, and his long-running email newsletter *Vital Weekly* regularly reviews tape releases. Ron Lessard's RRRecords is still open in Massachusetts, and you can still purchase titles from his *Recycled Music* cassette series. David Lichtenverg's tape-based radio show *Under the Floorboards* remains on the air in Denver, four decades after it began. Don Campau's program *No Pigeonholes* continued weekly for over thirty years, finally ending at the end of 2019; in addition, he put together a website called *The Living Archive of Underground Music* filled with photos, stories, and interviews from all kinds of figures in the underground.

Hal McGee has been just as busy, first with his print zines *Electronic Cottage* and *HalZine*, then with his website HalTapes, alongside the reams of old cassettes he sells in digital form on the online retail site Bandcamp. "I still have that same attitude of do-it-yourself and don't compromise," says McGee. "The music industry never conceived of blank cassettes as something for people to make their own music. They thought people were going to use them just to make copies of the albums that they bought from the big companies. But we used them for our own purposes, to create our own work and our own meanings. Life is basically meaningless except for the meaning that we invest in it. We're staring into an abyss, but art and friendship give it some meaning."

CHAPTER FOUR

INDEX
▶

The Tape Traders

A

Recording and Sharing Live Music on Cassette

Neil Corkindale was just nineteen years old, living in Manchester, England, when a judge in a high court branded him an "evil genius." What constituted this lad's cunning villainy? Selling cassette tapes of concerts he had recorded. The case against Corkindale was brought forth in 1978 by BPI; the teenager fled the country before the judge banned him from making or selling live tapes. But this was just one of over forty BPI suits against supposedly nefarious music "bootleggers." "Every time we go to the high court for an inspection order, we find ourselves acquiring new leads to fresh suspects," BPI director general Geoffrey Bridge told *Billboard*. "Often they may be retailers selling British or imported bootleg material, but we try always to trace the product back to the original source: the people who are actually making the tapes."

Lionel Mapleson holds up the kind of wax
cylinder he used to record concerts at
New York's Metropolitan Opera House.

In music, "bootleg" has long been a catchall term, used to describe
not just unauthorized concert recordings but also counterfeit copies
of official music releases, illicitly acquired studio outtakes and demo
material, and live broadcasts captured from radio and television. All of
these kinds of bootlegging started long before the compact cassette was
introduced. Within years of Edison's invention of the phonograph in the
1870s, people took advantage of lax copyright laws—designed to protect
performances of written music, before recording capabilities existed—to
make and sell copies of previously released commercial recordings.

Amateur live recordings started happening almost as early. Around
1900, Lionel Mapleson, an employee of New York's Metropolitan Opera
House, captured concerts there on his portable Bettini brand phonograph recorder. The device used wax cylinders that could capture two
minutes of audio and were reusable (though a disc's recorded grooves
had to be shaved down before each use, meaning multiple reuses eventually destroyed them). Mapleson made over 100 cylinders in three
years, then abruptly stopped. It's unclear why, though it was possibly by
request of Met officials, who hoped to sell their own official recordings
and did soon after.

In ensuing decades, both fans and profit seekers carted wire record-ers and reel-to-reel decks to performances to build personal libraries or underground businesses. The introduction of the cassette tape made this a lot easier. Unlike with expensive, complicated, bulky reel-to-reel gear, you didn't have to be rich to buy a cassette recorder, you didn't have to be a professional to operate one, and you could hide smaller ones under your seat or inside your clothes, reducing your chance of being caught. There was another bonus: just as cassette tapes offered an easy way to record shows, they also provided the chance to immedi-ately hear and share those recordings. Though previous formats could be duplicated and shared too, the difficulty and expense meant that most often, bootlegged concerts were pressed onto vinyl LPs. The music industry's argument against this—that it took potential earnings away from artists—was understandable. But it could also be argued that many people who bought unauthorized live albums had already purchased a given artist's official releases and would continue to do so, perhaps even boosting such sales through word of mouth.

Once cassette tapes became popular, the industry's pursuit of indi-vidual tapers became less defensible. That's because, starting in the late 1960s and then booming in the '70s, many "tape trader" subcultures often eschewed the idea of commerce altogether. They consisted pri-marily of megafans who wanted to hear everything by particular artists rather than opportunists seeking to cash in. They swapped cassettes through the mail or in person at shows and dubbing parties. They prized the documentation and sharing of music over its commodification and looked down on sellers. "I absolutely DISDAIN the word 'bootleg' because it implies evil intentions and the exchange of money," one tape trader told *Popular Music* in 2003. "I have NEVER charged anyone for tapes (either in the form of cash or extra blank tapes) and find it repulsive when others do. Tape trading is done because one enjoys music and wants to spread it to others; bootlegging occurs when one wants to make a quick, dishonest, buck."

Concerts have always offered a way for artists to connect with fans more directly, without a record company deciding what would be heard and when. So the attitude that live recordings are motivated by fan-dom rather than profit predates cassettes. Take Lionel Mapleson at the Met, who initially refused to license his recordings to commercial labels because, as he told the *New Yorker*, "they're much too personal." Record-ing live performances appealed to those fascinated with the continuing story that concerts told, as artists worked out their music in public in real time. As Alex Sayf Cummings points out in her book *Democracy of Sound*, this practice extended from Mapleson's time into ensuing decades, via "classical copiers of the 1950s and 1960s, who yearned to record the subtle nuances of each unique performance of a beloved opera singer . . . [and] jazz fans who valued every iteration of a composition by a virtuoso musician as worthy of preservation." In diary entries from his recording days, Mapleson sounds a lot like a modern taper: "For the present, I neither work properly nor eat nor sleep," he wrote. "I'm a phonograph maniac!"

By the late 1960s, phonograph maniacs had become tape maniacs, carving out their own live-recording communities via cassettes. Many genres had their share of taping devotees, including folk, jazz, opera, and bluegrass. Rock and roll was a particularly ripe scene for taping, as artists pushed against past constraints set by the three-minute radio-ready single, especially in concert. Many bands around the world were becoming more exploratory, more psychedelic, and more open to the unpredictability of improvisation. In this milieu, any given show could be unique, and tours could include frequent changes in song versions, set lists, and instrumentation. If you weren't there, you could miss moments that might never recur. Of course, all kinds of artists were being bootlegged in every sense of the word. But the most interesting, personal use of cassette tape arose when fans captured live music to share rather than sell.

The most famous and influential example of these burgeoning taper communities grew around a San Francisco band who formed in the mid-1960s, eventually becoming a subculture unto themselves. The Grateful Dead played a mélange of rock, folk, jazz, and more, and in concert they showed a predilection for long, improvised instrumental jams. In Michael Getz and John Dwork's series of *Deadhead's Taping Compendium* books, Dwork, a tape trader himself, described the band's music as "a catalyst that can spark in the mind's eye of the properly attuned listener an infinite variety of eternally unfolding, metamorphosing, multicolored architectures of shapes, contexts, and messages. Journeying through these infinitely intertwined and interrelated architectures, one can access the full spectrum of pure emotions, being states, and sensory experience." A bit hyperbolic, perhaps, but certainly a sentiment shared by many Grateful Dead aficionados, or Deadheads, as they came to be known. They found these transformative possibilities so enticing that they wanted to hear every iteration of the group's work. Each show was like a page in an ongoing, evolving musical diary—or as taper Dan Huper described to Dead historian Dennis McNally, "a compilation, every night, of every show that went before."

"I would see them four or five nights in a row and not see a single song repeated," says David Lemieux, a taper who became the Dead's official archivist. "Maybe one night they busted out a song they hadn't played in ten years, and that made that show extra special." "What they did on any night would be unique," adds Robert Wagner, a doctor who got hooked on the Dead as a college student in the 1970s. "Some nights were more inspired than others, but they're all good in their own way, even when they're not at their best. So you want to hear them all, and the only way to do that is to have the tapes."

The tape-trading scene that formed around the Dead began slowly. In the 1960s, few people thought to record the band from the audience. Those resourceful enough to try had to afford, operate, and transport expensive, heavy reel-to-reel decks. As a result, according to some tapers, fewer than thirty audience-recorded tapes from before 1970 have shown up in trading circles. "Finding those early tapes back then was sort of like finding the Dead Sea Scrolls," taper Eddie Claridge explained in the *Compendium*. The Dead's own sound engineers did record some shows themselves, though, and copies of these recordings occasionally made their way out to the fans.

In the 1970s, Dead show taping began to gain steam thanks in part to high-end cassette recorders. The Nakamichi 550 in particular became a fan favorite, even though it was still rather large and heavy. Traders formed clubs to build their collections of both audience and sound-board tapes. One of the earliest, the First Free Underground Grateful Dead Tape Exchange, was started by Les Kippel in 1972. It soon spawned another eight clubs just in Kippel's own city of New York and upward of thirty more around the country. "A lot of people want to set up exchanges," Kippel told *Rolling Stone* in 1973. "I tell them to get cards made up with their telephone numbers on them, but I also insist it says 'free' on them."

That *Rolling Stone* article, which dubbed Kippel "Mr. Tapes" and boasted about his over-500-hour collection, flooded him with tape requests. "We would have taping sessions where people would come to my house and we'd run tape machine to tape machine to tape machine," he recalled in the *Compendium*. "The peak of absurdity was when I had thirteen tape machines running at once! Thirteen people there, 'Okay . . . Ready, set, put your fingers on the pause button and—go!'" One Kippel disciple, John Orlando, even claimed to have quit his day job so he could spend eighteen hours a day dubbing Dead tapes. "I had to stay high all day," he told *Rolling Stone*. "Or I'd go nuts."

In 1974, Kippel started a magazine called *Dead Relix* that spread information about show taping and trading. Articles included advice on how to record, what kinds of decks and tapes to use, and proper trader decorum. There were also sections where fans could list recordings they were willing to trade (mostly by date and venue) or sought to trade for. Usually all that was required to tap into these personal libraries were a few blank cassette tapes and a little money—strictly for postage, not profit. In the first issue of *Dead Relix*, an editorial insisted that they "in no way advocate the duplication of live recordings for purposes other than free exchange. . . . Tape trading is based on HONESTY!"

Dead Relix helped spark a mid-1970s boom in Grateful Dead taping and trading. So did the introduction of cheaper, smaller cassette recorders with decent-quality audio, such as Sony's portable 152SD, which included professional-style level indicators known as VU (volume unit) meters. Another boost came, oddly enough, from the Dead's absence. When the band went on hiatus in 1975, old and new traders had time to fill out their collections rather than chase down the latest tapes. "When I started [college] in 1975, I probably had about five Dead tapes," said taper Mark Mattson in the *Compendium*. "By the time I got my B.A., I had to move almost a thousand tapes." Once the Dead returned to touring in 1976, their fan base had actually increased due to tape trading, and there were enough tapers to cover every concert. It's likely that every show for the rest of the band's existence was recorded by at least one member of the audience.

The community was also boosted by more fans creating clubs and magazines dedicated to Dead tape trading. Colleges in particular were a hotbed. In 1979, at Hampshire College in Massachusetts, John Dwork started a Grateful Dead Historical Society and an in-house newsletter called *DeadBeat*. After graduating with perhaps the first ever undergraduate degree in Frisbee (or, officially, Flying Disc Entertainment and Education) he started the *Terrapin Flyer*, which later morphed into the full-fledged magazine *Dupree's Diamond News*. Alongside *In Concert*

Quarterly, Golden Road, and others, it offered a wealth of Dead concert information, reviews, taping tips, and want lists.

By the 1980s, anyone even remotely interested in the Dead beyond their official studio recordings had endless opportunities to spark up a tape addiction. As *OP* magazine did in the cassette underground, and metal fanzines did with band demos, Dead-themed publications linked fans to bands and one another, fostering communities that grew exponentially. "I was living in my mom's house, and virtually every day an envelope or a box of tapes arrived," remembers David Lemieux. "It was as exciting as Christmas morning." Swaps happened in person too: trader Paul Scotton recalls taping parties where "if you showed up with cords and a deck, you could plug in and we'd do it until we couldn't stand each other anymore. It could literally go on for days."

As more people entered the Dead taping scene, some resentments festered between veterans and newcomers. "Some people, though not all of them, were very protective; they viewed their tapes as intellectual property and weren't going to trade with just anybody," explains Robert Wagner. "But when I became a taper myself, I gave copies to anyone who wanted them. People would come up to me at a show and ask if they could plug their deck into mine, and I'd say, 'Sure!'" "I think it definitely was part of a closed club, but it was something that most of us loved to share," agrees Lemieux. "We didn't feel superior, certainly not elitist, just different. There were so many different subsets of Deadheads, and tapers were just one of them, no more or no less important than any of the others."

Still, it was understandable that some tapers would be protective, given how much work it could take to make a tape. Before their 1975 hiatus, the members of the Grateful Dead were somewhat resistant to audience recording. In the *Compendium,* Kippel recalls sitting in a hotel room after a 1972 show, listening to a tape he had made with friends, when suddenly the Dead's sound engineer Owsley Stanley busted in, ripped the cassette from Kippel's deck, and stormed back out. Others tell stories of Dead roadies cutting their microphone cables when they were discovered taping. In response, tapers devised all kinds of stealth ways to record shows. At first it wasn't too tough to get past a concert venue's security staff, who often weren't even aware of audience taping and could easily miss a deck buried in a bag or believe a taper who said their gear wasn't for recording. Some tapers went to great lengths to hide their work, spreading gear among different attendees as they arrived, or stuffing entire mic stands down pant legs and claiming to be disabled. Once inside, tapers hid decks under seats or coats, only pulling them out when the lights went down, or venturing into bathroom stalls to assemble their recording setups. "It's a real pain in the ass doing an audience tape," Kippel told *Rolling Stone*, listing the many preparations required, including "a lot of dope to keep you mellow."

Eventually, the Dead warmed up to audience taping. When queried in 1977 about the phenomenon, bandleader Jerry Garcia at first bemoaned unauthorized recordings, until he realized the interviewer was talking about trading live tapes, not selling bootleg albums. "Oh, [live] tapes I don't care about—the tapes are totally cool," he said. "I spend a lotta time in bluegrass music doing the same kind of stuff: swapping tapes and doing all that." (Garcia had learned some bluegrass-style guitar playing by listening to live tapes he got from collector Marvin Hedrick; he even taped some bluegrass shows himself.) "We could either let them come in and tape and take it with them, or we could become cops and take away their machines," drummer Mickey Hart told *Billboard* decades later. "We had a meeting and said, 'We don't want to be cops!' So we let them do it." By the early 1980s, tapers could set up their gear out in the open at Dead shows without hassle from the band's crew, even when the lights went up.

Unfortunately, security at venues had gotten wiser about audience taping and was less lenient. Even though Dead sound engineer Dan Healy would tell guards that the band didn't mind, tapers still often had to hide their gear in clothes, bags, and even wheelchairs to safely enter shows with everything they needed to record. Taper Frank Streeter remembered bringing his gear piecemeal in the clothes of five different attendees, who would meet in the bathroom, disrobe, and assemble it all. At one venue, taper Chris Hecht left his gear outside under a bathroom window, then pulled it up with a rope once he was inside. Some smuggled mics inside of shampoo bottles or sandwiches; others buried gear under floors or hid it in janitor's closets between shows during multinight Dead stands. It made things difficult, but it also kept the underground nature of audience taping strong. This was a community operating outside normal channels, using cassettes to circumvent the system.

It was also a community that turned show taping into an art form. All the decisions that go into making a tape—what gear to use, where to stand in the venue, how to place microphones—came to reflect the personality of each taper and colored the experience of trading and listening. "It becomes a creative art, which I didn't necessarily anticipate when I started doing this; I just wanted the music," says Robert Wagner, who started taping shows himself in the late 1970s when he was frustrated with the quality of tapes he got via trades. "But eventually my ear became trained, and I learned from other tapers to listen to each tape you make and try to make it as good as you can."

There were also craft skills that came with experience. "You want to know where you were in the set, and where the next break is coming so you can flip the tape and not cut off a song," says Paul Scotton. "You can only know that by going to a lot of shows and making a lot of tapes." As Dwork put it in typically dramatic fashion in the *Compendium*, "We have mastered the martial art of equipment sneaking, tape flipping at the speed of light, superior ticket ordering, insane tour scheduling, even sending our recording decks on tour while we stay at home. We've developed our own language, our own culture, and, for some, even our own spiritual dimension!"

There was also an art to the archiving and documentation of Dead tapes. Each taper had their own approach to labeling, sometimes adding details of show dates, song titles, gear used, and where mics were placed. Some traders were happy to have many different label styles adorning their racks; others insisted that tapes be sent back to them with the covers still blank, so they could add information in their own preferred manner. Perhaps unsurprisingly, many were so detail oriented that they could be seen, as Wagner puts it, as "hyperfocused savants." "If I taped a run of, say, twelve shows, I could remember where I was at each one," he recalls. "I could rattle off the top of my head the set list for every one. I would know that they went from one song to another then back to that first song and so forth." Wagner's obsessions were rewarded when he made a last-minute decision to drive to a Mississippi gig in 1978. He turned out to be the only audience member there who was taping. Later, he discovered the Dead's own recording didn't survive, so he owned the only document of that show. Originally drawn in by the uniqueness of every Dead concert, Wagner now had something unique of his own to contribute to their vaults.

For the first time, tapers set up their microphones and decks in a Grateful Dead–approved section at a concert on October 27, 1984. (Photo by David Gans)

As tapers continued to refine their art and the Dead continued to accept them, it became common to see a nest of microphones thrust up high in front of the soundboard. The taper thatch grew so large that eventually sound engineer Dan Healy had trouble seeing the stage as he mixed the band. He also got frustrated hearing stories of tapers stealing seats from nontaping fans by claiming they got there first to set up all their gear. In 1984, during a three-show Dead stint in Berkeley, California, Healy unveiled a solution to these problems: a dedicated section of seats behind the soundboard for tapers, for which tickets could be purchased ahead of time through the mail. The idea brought some immediate advantages: tapers could bring in lots of gear with ease and could more quickly share recordings, sometimes chaining their decks together to access one common audio signal.

But there were downsides too. Veteran tapers complained that the section drained some of the fun out of the experience, with all the equipment blocking movement and tapers policing people in the section to stay quiet. Even worse, some tapers felt the sound quality of what they captured from behind the board was inferior to what they got when closer to the stage. "To sit in the tapers' section and get a terrible recording wasn't even worth going to a show," taper Dougal Donaldson said in the *Compendium*. "To ghettoize all tapers like that was almost a punishment." In response, many tapers ventured back to their previously preferred spots, resurrecting old ways to record without being noticed. It was perhaps a blessing in disguise, allowing the community to maintain the underground aspect of its operations even as the taping phenomenon became well known.

Whatever the benefits and detriments, the Dead's official acceptance of taping publicly acknowledged a crucial part of their fan base. It signaled to tapers, and anyone else interested in the band, that the group appreciated its most devout followers. "The audience is [as] much the band as the band is the audience," drummer Bill Kreutzmann told Dead historian Dennis McNally. "There is no difference. The audience should be paid—they contribute so much." "It was a very smart move on the Dead's part," says Scotton. "All this trading made people more and more interested, and they realized it wasn't the same show over and over, so they started packing halls even more." "I also think the Dead firmly knew that the strength of their cultural identity was in their live performances, and it was very hard to 'get' the Grateful Dead from their studio records," says Lemieux. "I love the studio records, but the power, the thing that would turn you on and literally change your life was hearing the Dead live in concert. And the second-best thing to that was hearing a live tape."

Into the late 1980s, the Dead's popularity certainly grew, whatever the reason. Despite not having hit records (save for a lone 1987 appearance in the *Billboard* Top 10 by a studio recording of the live staple "Touch of Grey"), they became one of the most financially successful touring bands around, taking in $50 million a year from shows during their peak. They sustained a large audience regardless of how often they released studio albums—which, unlike with many other bands, the tours weren't really meant to promote anyway.

Throughout all these years, soundboard tapes continued to circulate. Some tapers sought the best-generation versions of those recordings; later, after the advent of digital recording, some mixed soundboard recordings with audience tapes for the best combination of musical quality and ambience. For many, audience tapes offered a more "real" version of a show, less sanitized than the sound of the instruments sent directly to the mixer from the band's onstage mics and amps. "A sound-board tape might be perfectly balanced, but it often lacked something," says Wagner. "The audience tapes made you feel more like you were there, with the crowd noise and hall ambience. Under good conditions, that added life to the tape."

In the 1990s, the Dead began releasing official soundboard record-ings in a variety of forms. One series was begun by Dick Latvala, a for-mer tape trader who in the 1970s amassed a collection of 800 shows on reel-to-reel. After he became the band's official archivist, he launched *Dick's Picks*, periodically releasing concert recordings from the band's vast vaults. When Jerry Garcia died in 1995, the most famous era of the Grateful Dead ended (though other members have since reunited under other group names). But *Dick's Picks* kept going. When Latvala passed away in 1999, David Lemieux took over. Aside from being a rabid trader himself, Lemieux had a graduate degree in film preservation. He had originally contacted the Dead to ask if they had a video archive of their concerts. The band hired him to help catalog that section of their vault, making him a natural choice to take over the entire Dead library after Latvala died.

Lemieux continued *Dick's Picks* until 2005 and launched his own *Dave's Picks* series in 2012. By then, the band and most tapers had moved away from cassette, using digital audio tape, CD, and other for-mats as early as the 1990s. But Lemieux still uses the band's own sound-board cassette recordings for some older releases in his series, and he's often impressed by how well the tapes have held up. "I go back and I listen to my [digital audio tapes] from the early nineties, and they have tons of problems, dropouts and so forth," he says. "Whereas when I put on any of my cassettes from the late eighties, they sound as good today as the day they were recorded."

Long after cassettes faded from the scene, the Dead trader com-munity remained strong, and many friendships forged at concerts and through the mail continue to endure. "That was part of the experience of going to a show, seeing old family and friends, running into people you hadn't seen in thirty shows, and all of a sudden there they are again," Scotton remembers. "When you listen to the music, you remember those times. The fact that we can pull this stuff out from tapes, I think it really matters. It sends chills down my spine when I hear it. There are times when space and time don't matter, you're just off somewhere—and these guys were there, just night after night after night. It's remarkable."

Though few spurred a scene as big as that of the Grateful Dead, countless other musical artists were recorded on cassette by fans from the 1960s on. One community nearly as obsessed and devoted as the Dead's followers grew around the work of Bob Dylan. "Their aim was to be complete, to own everything, every concert, every studio outtake, every known tape floating around anywhere in the world," wrote David Kinney in his book *The Dylanologists*. "Some completists were not satisfied until they owned every taper's recording of a particular concert."

Dylan tapers went to great lengths to record shows, attaching gear to legs and backs, stuffing it in pillows under shirts, even hiding it inside hollowed-out loaves of bread and half-full thermoses. Competition was fierce. Some were rumored to own tapes that they would never show or even mention to anyone else. Some lied about wanting to just hear a rare tape, then surreptitiously recorded it with a device in their pocket. Ultimately, though, most Dylan collectors prioritized the music. "The tapers believed they were doing important work," wrote Kinney. "They were documenting music that otherwise would have floated off into the ether, never to be heard again."

Brigid Berlin knew a lot about saving things that might disappear. As part of Andy Warhol's Factory scene in New York, she made a habit of carrying a tape recorder, capturing the sounds of everything happening around her. One night in August 1970, she brought her Sony T120 to the nightclub Max's Kansas City, where the Velvet Underground had a summer-long residency. She recorded their set that evening on a ninety-minute Norelco brand tape, which read on the front, "LIFETIME GUARANTEE." The audio is raw and noisy, with lots of audience chatter. Between songs you can hear writer Jim Carroll, who often held Berlin's microphone, talk about getting drinks from the bar. For Berlin, this was just another night of taping, but it turned out to be Velvet Underground founder Lou Reed's last show with the group. Danny Fields from their label, Atlantic, asked Berlin for a copy and released selections in 1972 as *Live at Max's Kansas City* (a later reissue included the entire tape). At the time, *Rolling Stone* called it "in some ways the first authorized bootleg." Though many tapes of Velvet Underground gigs would later surface, *Live at Max's Kansas City* remains one of their most well-known live albums.

No such live document emerged during the career of the German group Can, which is surprising, as they were more given to live improvisation and night-by-night variance than perhaps even the Grateful

Dead. Fortunately, in 1973 a British fan named Andrew Hall took a bulky Sony tape recorder to a Can gig in London. "[It] could only be contained within some size 36 trousers (I took size 28 at that time), a similarly oversized shirt and a hefty duffel coat over that," he wrote in liner notes for the recording's release decades later. "If the temperature was turned up . . . I just about melted." He continued this practice for years and was eventually invited by the band to stand next to the sound-board while he worked. In 2021, Can released *Live in Stuttgart 1975* and *Live in Brighton 1975* from Hall's tapes, with plans to make more of his recordings available in the future.

Following the Grateful Dead's example, some bands formally adopted an open policy toward audience recording. Metallica even approached the Dead for advice on setting up a taper area. The Dead showed them the standard documents they gave to promoters and fans, and Metallica copied them—with the band's approval—word for word. Phish, a jam band from Vermont and the Dead's closest stylistic descendant, also instituted a space for taping at shows. The advent of the internet, par-ticularly message boards, helped create a community around Phish tape trading that's still going strong.

While attending high school in New Jersey, Paul Nixon found that community on an internet bulletin board hosted by the company Prod-igy. He soon had to convince his parents it was safe to send and receive tapes from strangers through the mail. "Northeastern New Jersey was a pretty square place, and I wanted to find the weirdos," he remembers. "This Phish thing seemed like a good lead to follow." Inspired by the way no two Phish shows were the same, Nixon dove in. "Phish became about community for me," he says. "There were familiar faces, a safe atmosphere, and despite a few super-weirdos, usually only good times. When I finally met the main guy I traded with on Prodigy in real life at Madison Square Garden, it was like we were already friends, and we hugged and it was all good."

Despite the good graces of welcoming bands such as Phish, recording concerts has remained a dangerous game. In their anecdotal 2011 survey of tapers, Mark Neumann and Timothy Simpson relate a wide range of stealth stories. Their interviewees have dispersed gear among a group, put illicit substances such as alcohol in the same bag as tapes to distract security, and even donned outlandish costumes that they claim are effec-tive. "You go with a girl," said a taper who goes by the name Satch. "Take an old basketball that's deflated and has a hole in it or whatever, and cut it in half. Put it underneath her like she's pregnant, and [have her] wear a big maternity shirt or something. You can get a video and audio recorder in [that way]."

Among show tapers and traders who have persisted despite these obstacles, no two do it in exactly the same way. The most devoted can be like scenes unto themselves, developing their own methods of capture, organization, dissemination, and communication. Nearly all of them (at least among those who don't do it for profit) are rabid fans and documentarians, and usually big proponents of the cassette tape.

In the 1970s, New Yorker Lee Ranaldo frequented Grateful Dead shows and collected live tapes, mostly on reel-to-reel. He also sought concert recordings by the Beatles, Neil Young, the Velvet Underground, and Bob Dylan. "It was about wanting more from these artists, about not being able to get your fill," he remembers. "These were artists who were not trying to reproduce songs exactly the same every night but rather be varied enough so that each show was new." Ranaldo's evolution into a tape obsessive clicked when he heard a Dylan recording from a mid-1960s European tour. An acoustic version of the song "Visions of Johanna" fascinated him so much that he decided to track down every Dylan tape he could. He became particularly passionate about *The Basement Tapes*, which he found tapes of long before Dylan put out an authorized edition. Once that official version came out, "I gathered all of the takes from two different bootlegs and the official version, and I created a crazy spreadsheet going back and forth across them all," he remembers. "I wanted to find the best version of every single song, because sometimes the bootlegs were better than the official version. This is where tapers get really obsessive. It was one of those rabbit holes." (Ranaldo has since helped with research at the Dylan Archive in Tulsa, Oklahoma.)

In the early 1980s, Ranaldo formed the band Sonic Youth with Kim Gordon and Thurston Moore, and their fascination with tapes heavily influenced their approach to music. During a 1982 tour, they taped their performance each night, then listened to it on the way to the next gig. "We were sort of taking them apart and figuring out what we did," Ranaldo says. "It was this rapid, repetitive thing. You're inside this feedback loop of playing the songs, listening to your versions, playing them again, listening to those versions. They couldn't help but inform each other. That led to us being obsessive about taping everything we did, from the earliest days of the band." Not long after, Sonic Youth's fans began taping their shows too, which the band welcomed—especially if they got a copy for themselves. Sometimes, those fan-made tapes were used as source material for official live albums. Unofficial bootlegs of shows also popped up on vinyl and CD, bringing Ranaldo's taper life full circle. "That completely amazed us," Ranaldo remembers. "Here we were, music fans who have collected bootleg albums ourselves, and all of a sudden, we were being bootlegged! There was something really cool about that—like, wow, we're in the pantheon now!"

Like Ranaldo, Robert O'Haire began attending concerts in New York in the 1970s and quickly became obsessed. In 1981, he borrowed a small cassette recorder from his dad—who used it in his reporting for the *New York Times*—and took it to a Frank Zappa show at Long Island University, expecting to get a recording as good as the official live albums he owned. "I listened to the tape, and I was horrified to hear myself," he remembers. "All I had managed to get was me and my friend talking and screaming throughout the entire set." Luckily, O'Haire's mother bought him a higher-end cassette recorder with which he could raise and lower recording levels as needed. Once he began taping regularly, he devised methods to avoid getting caught. He carried extra blank tapes in his pockets so he could start recording again if his first one got confiscated; he stuck batteries in his socks in case security insisted on ripping the originals from his recorder as he entered a venue. "I'd have to try to minimize how much I touched my recorder," he explains. "Because they'd be spying on me the whole time."

Along the way, O'Haire caught some indelible moments, such as the final performance of Johnny Thunders's Heartbreakers before the punk icon passed away, and an early 1990s Cramps show in which latex-clad frontman Lux Interior "roasted like a sausage" under spotlights. "The serotonin is running through your body, and it's just the best," he says of such experiences. "It's all in that moment." The moment is particularly important in improvised jazz, the kind of music that O'Haire has recorded the most, including more than fifty concerts involving pianist Cecil Taylor. His library contains over 1,000 recordings by over 100 artists and includes ephemera such as photos, ticket stubs, newspaper ads, and set lists. The work helped spur his career as a sound recordist and filmmaker, after musicians noticed his techniques and offered to pay him. "A lot of these shows are from small, DIY venues," O'Haire adds. "Imagine if we had all the recordings from these cool little places that churned out amazing music. When it's gone, it's gone. It's in the air."

Aided by the postal system and the internet, rabid tapers and traders can also be found outside of big cities such as New York. Growing up in York, Pennsylvania, in the 1980s, Henry Owings discovered live tapes as a lifeline to places beyond his small town. "I was lonely," he admits. "I was trying to reach out to the world and figure stuff out." Traveling to cities along the East Coast, Owings taped shows and connected with traders, finally hearing bands he had only read about. Sometimes when getting a tape in the mail, he would spot an address crossed out on the envelope and blindly send a letter there to see if that person was into trading too (usually, they were). "With a lot of the bands I liked, if you had a cassette of one of their shows, it was like you were instantly part of a club," he remembers.

Owings recorded and traded tapes of early shows by bands such as Seattle's Nirvana, DC's Fugazi, and Ohio's Bitch Magnet. He guesses that he probably has more live tapes by Chicago's Jesus Lizard than they themselves do. "I recorded everything, and I still do—it's like a fucking curse," he says with a laugh. But he doesn't let his obsession with documentation get in the way of the music. "I like to move around, so I would just set up my microphone and walk away," he remembers. "That's very counter to what I see a lot of recording people do now, where they just stand there with their gear. I'm just like, 'Dude, that seems like the most boring way to enjoy a show.'"

Owings became an archivist too, digitizing his tape collection to build a mini-history of 1980s and '90s underground American rock. Much of it is now housed in the special collections department at the University of Georgia, an archive to which Owings continually adds. (Owings also documented this scene through his irreverent magazine—and record label—*Chunklet*.) For Owings, live cassettes may be history, but the community he found is still around. "I have friends to this day that I met in the back pages of *Maximum Rocknroll* and *Flipside*," he says. "We're all freaks, and this was how we scratched that itch. Recording and trading gave me an outlet for all this energy."

If any underground rock taper has been more obsessive than Owings, it's Aadam Jacobs. As a Chicago high schooler in the early 1980s, he first captured a concert with a recorder borrowed from his grandmother. Soon he upgraded to a Walkman with a built-in mic, and later actually hauled his big component tape deck to shows, sometimes carrying it in a suitcase. "The passion started with having unique recordings by favorite bands, but it soon led me to have a reason to talk with them and make

A selection of Henry Owings's cassette recordings of concerts from the 1990s. (Photo by Henry Owings)

me stand out as a fan," says Jacobs, who would sometimes bring two decks so he could make a tape for the band too. "It soon became difficult to enjoy a show without recording it."

Throughout the ensuing decades, Jacobs could be found with his microphone raised at seemingly every show in Chicago. His main haunts included the West End, the Lounge Ax, the Empty Bottle, and the Metro, all of which welcomed his taping presence (though some run-ins at the Metro got him temporarily banned). Sometimes he would record shows at two venues in the same night, riding between them on his bike with his gear strapped to his back. Often he would patch into the club's soundboard as well as tape with his own recorder, then later mix them together. "I stopped counting in 2000 when I was at about 6,000 shows," he says. "I suppose I'm near 10,000 now."

Jacobs became so well known among fans and bands that the latter would sometimes request his taping services. In 2002, veteran punks Mekons asked him to record their twenty-fifth-anniversary show and released part of it on a subsequent album. But Jacobs's habit was never about fame or fortune. It was a fixation he couldn't quit, so consuming that his apartment full of cassettes left no space for visitors. "It did dominate my life," he admits. "But I was having fun, and l look back on it with few regrets."

Music has dominated Pete Gershon's life since he was an undergrad in Massachusetts in the early 1990s, listening to jam bands such as Phish, Aquarium Rescue Unit, and Widespread Panic. At first Gershon traded tapes of all these groups, but "when I figured out that a lot of these bands let you record their shows, I started arriving with a bulky, ancient Sanyo cassette deck that required AC power," he remembers. "Most of the bands I liked excelled on stage, brought on special guests all the time, and made improvisation central to their practice, so you would always hear something different from show to show."

Moving to Vermont after graduation, Gershon went to four or five shows a week, carting along his portable Sony D-6 tape recorder and Shure 737A microphone. Sometimes he plugged into sound-boards; other times he held his mic as close as he could to the stage. "I already had a couple hundred first- or second-generation tapes when I moved," he remembers. "Now I was stacking up 'master tapes,' usually on kind-of-expensive Maxell XL-II blanks, at least a dozen or two every month." Once some of these bands started to play outside of Vermont, Gershon got frequent requests for copies from his growing library.

In the late 1990s, Gershon launched *Signal to Noise*, a magazine covering much of the music he had documented for the past decade. That took time away from his taping habits, but he still amassed another few hundred gigs, continuing after *Signal to Noise* shuttered in 2013. "You used to really need to make an effort to buy and maintain equipment and recording media, get to the gig early so you could set up your mics, and then catalog and copy recordings afterward," he says. "The work of a few individuals can make the difference as far as ensuring a lasting legacy for creative people. [I liked] being part of a tradition of people who make the time and effort to capture and preserve a creative moment that would easily be lost otherwise."

Capturing moments was also a prime motivation for Pieter Schoolwerth, who discovered the 1980s cassette underground as a teenager in Richmond, Virginia. "It was inspiring to see hundreds of artists totally bypassing the entire commercial music industry," he remembers. "They all had dashed-off mail-order catalogs clinically listing their releases in a typewriter font for three to five dollars each, and they often generously took the time to handwrite effusively friendly letters back suggesting lists of tapes by other freaks in the middle of nowhere." Schoolwerth also noticed ads in *OPtion* that offered concert recordings for a little cash and a self-addressed stamped envelope. When one trader from England sent him ten ninety-minute tapes of shows from all over Europe, he was hooked.

After saving up money to buy a Walkman that could record, Schoolwerth started taping shows himself and quickly built up a list to send to trader friends. "Some of us had names for our 'anti-labels,' but there was an unspoken understanding that once you received a tape in trade you were welcome to copy it and relist it in your collection," he explains. "Most of us knew where the tapes originated, and a few people made their own cover art with a picture and oddball anecdotes scrawled on to denote the spirit of that particular gig."

In the late 1980s, Schoolwerth got hooked by two groups known for their unpredictable live sets, Psychic TV (featuring Genesis P-Orridge of Throbbing Gristle) and Butthole Surfers. "I loved recording their gigs, as the set lists were flexible and always left a nice black hole of space each night for improvisation," he recalls. "With Gibby [Haynes of Butthole Surfers] and Gen rambling on stream-of-consciousness-style between songs, it seemed anything could happen." He also continued to collect live tapes through mail trades, becoming especially intrigued by "bad quality" recordings. "[Those are] tapes where the recording was all fucked up in some way that allowed you to feel the atmosphere of the room and scene beyond the band's performance," continues Schoolwerth, now a respected visual artist. "These unexpected international auditory aberrations I gathered were a direct gateway into what it was like to be there."

Eric Hardiman had just graduated from college in 1990 when he came upon a musical gateway of his own. It was the work of Sun Ra, the visionary pianist, composer, and bandleader who claimed to be from

A portion of Pieter Schoolwerth's concert recordings made on cassette tape in the 1980s. (Photo by Pieter Schoolwerth)

Saturn (which is why he discarded his birth name, Herman Blount). Ra's work ranged from blues to bop to wildly adventurous improvised jazz with cosmic overtones, augmented by the glittery regalia in which he and his group the Arkestra performed. Not long after Ra passed away, Hardiman was living in Washington, DC, and listening to public radio station WPFW, which was playing several days' worth of nonstop Sun Ra material, including many live recordings. "I remember rushing to the store to buy as many blank cassettes as I could afford," Hardiman says. "I sat by my boom box and taped everything, setting my alarm clock to wake me up every forty-five minutes for side changes. These tapes became my constant companions as I walked around DC that summer. The more I listened, the more obsessed I became."

Soon Hardiman found an email list dedicated to Sun Ra enthusiasts who mentioned trading live tapes. He offered up his WPFW recordings, and suddenly cassettes were arriving at his doorstep weekly. "My living room filled up with blank tapes, printed lists of other traders' shows—sometimes a dozen pages in tiny single-spaced font—and cardboard mailers," he remembers. "I can still feel the excitement of hitting play on some of those tapes, closing my eyes and imagining what it must have been like at those Arkestra shows." Hardiman soon figured out which traders had the best-quality dubs, as well as which ones would toss in something extra, like photocopied show flyers or homemade stickers adorning tape covers. One trader even sent him a program saved from Sun Ra's memorial service. Much like the Grateful Dead, Sun Ra inspired this kind of obsession because his music was always changing. Hardiman's collection includes solo shows, duos, trios, and large ensembles with over twenty musicians on stage. Some tapes contain private Arkestra rehearsals at Sun Ra's compound in Philadelphia, where the entire group lived. And each tape tells a story, such as a tape of a 1971 show in Los Angeles where the power shut off, causing Ra to cast a curse on the city.

"All of the tapes added up to puzzle pieces for me, offering one more tidbit of information in the weird and wonderful mythology of Ra," says Hardiman. "The fact that someone else had prepared each tape for me only added to the depth of the experience." That led him to start his own cassette label, Tape Drift, in the mid-2000s. "I'm still buying blank tapes, duplicating piles of them, and sending them around to customers and friends," he says. "The connections I built in trading tapes have led to numerous real-world friendships that are vitally important to me."

In the 1990s and 2000s, as more people turned to CDs and then digital files, concert taping and trading happened less on cassette tape. But some from later generations have carried the practice forward, particularly when it comes to the Grateful Dead. In the mid-1990s, when Brian Weitz was in his early teens, a classmate told him a secret. A school staffer had a trove of Dead tapes and was willing to share. Weitz, who had already heard some live Dead recordings via his cousin, was intrigued. "I remember walking down to the kindergarten classrooms and waiting for him to have a break," Weitz recalls. "I was like, 'Hey, I heard you made tapes for people.'" Next thing he knew, Weitz was handing his connection blank tapes and getting Dead shows in return.

In high school, Weitz and his friend Dave Portner found a similar band to become obsessed with: Pavement. The group's punk-influenced, irony-tinted music became a college radio staple and didn't sound much like the Grateful Dead. Yet the bands did have one thing in common. "They played a lot of songs in concert that hadn't made it to albums yet," Weitz remembers. "Seeing them in 1994, right when [their album]

Crooked Rain, Crooked Rain came out, we heard a bunch of songs not on that record that ended up on the next one." Portner soon discovered Pavement fans on the internet—"There's a Grateful Dead world around Pavement that we didn't know about!" he told Weitz—and the two started trading live tapes with people they met online.

For Weitz and Portner, diving into Dead and Pavement tapes had a lasting impact, especially on Animal Collective, the band they formed with friends Noah Lennox and Josh Dibb. Their mix of noise, folk, and art-punk often expands in concert, where they blend songs through jammy interludes and regularly perform unreleased material. They even managed to include an authorized Grateful Dead sample in a song, adding parts of "Unbroken Chain" to their 2010 track "What Would I Want? Sky." But the Dead's biggest influence on Animal Collective is probably their attitude toward audience recordings. "We've always had an open policy, and we still do," he says. "Our manager tells every club, 'Don't confiscate recording equipment if you see it.'" As a result, a community of show traders has formed around Animal Collective, mirroring Weitz's and Portner's own experiences—even if their shows are traded via digital files now rather than on tape.

Growing up in the 1990s in western New York, Darryl Norsen didn't get to see any version of the Grateful Dead until after Jerry Garcia died. His parents finally let him go to a show when he was sixteen, and soon after, he gravitated toward internet forums dedicated to the Dead and Phish, discovering tape trading. When he tracked down some 1970s-era Dead shows that had happened near his home, something clicked. "I went from having twenty tapes to 200 within a year," he says. "People have various addictions. I think my addiction pretty quickly was accumulating music."

In the 2000s, as more and more people discarded their Dead tape collections, Norsen began grabbing them from those either giving them away or selling them for cheap. Meeting these fans was as interesting for Norsen as getting the tapes, since they often told him stories about their experiences with the band. "There's a lot of history in these tapes and the people that you meet," he says. "That's what made me want to collect more. I thought, 'How do you collect the tapes but also collect the stories?'" Norsen even went through a few cycles of giving away tapes himself. When he moved to Boston in 2003, strapped for space, he donated his entire collection to a record store. "I look back at that now and think, 'That was stupid!'" he says. "But it perpetuated the karmic reasons of doing this. By putting it out there, you get more in return." In the mid-2000s, he wrote a Dead-themed column for the online magazine *Aquarium Drunkard,* and the collector bug bit again. Again seeking out those shedding themselves of tapes, he took just six months to build a 5,000-tape library.

Norsen's urge to share took over again after seeing a memorable Dead-related show in 2018. "I felt like I experienced what every Deadhead is trying to chase: the perfect show, perfect moment," he says. "I came to the realization that in owning so many tapes, I was trying to fill that story, that hole for myself, and it wasn't as important as witnessing and experiencing it." He once again gave away tapes, hoping that his experience would inspire others toward Dead epiphanies. "I realized it's just dumb to hoard music," says Norsen, now an in-demand designer of album covers and associated artwork. "Music should be shared. It shouldn't be sitting on a shelf."

During his Dead tape travels, Norsen met a fellow collector named Mark Rodriguez, and the pair often had long phone chats, trading stories and Dead lore. Rodriguez had briefly collected Dead tapes as a high schooler in the late 1990s, but drifted away from it after graduation. When his interest in the band was rekindled a decade later, he had become an artist, and he started thinking about all the tapes that must still be out there. "I thought, 'No one's doing anything with these tapes,'" he remembers. "There's no value to them other than the quality of the music on them, which you can find online now. And unless someone actually has a tape player and is still listening to their collection, probably not everyone even wants to keep tapes. Why would you want to take up space with them?"

To find out, Rodriguez went on a quest. He first put an ad on Craigslist while living in Los Angeles and got a response from someone willing to hand over his collection of 100 Dead tapes. Soon, he connected with similar people all over, taking their cassettes as well as set lists and other ephemera. "At points there were, say, three collections I would pick up in one trip, so I'd have a trunk full of thousands of tapes, and I'd talk to these people and hang out with them," he says. "I started seeing each tape as this kind of charged object that not only had the sound of a show but also this value that someone ingrained in it since they first made or traded for that tape. Sometimes it was like this whole person was encapsulated in all these little tapes."

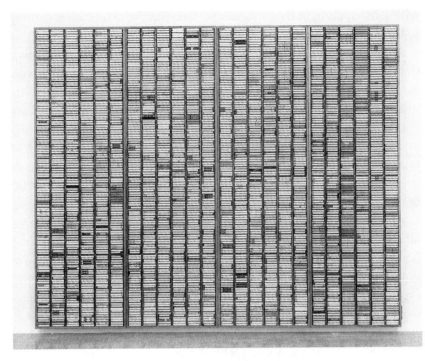

Mark A. Rodriguez, *1st Gen*, 2010–16.
Mahogany shelving units, 2,976 audio
cassettes of Grateful Dead live performances
in chronological order from 1965 to 1995,
plastic tape cases, ink and marker on paper,
oil-based enamel, 95.25 × 118.5 × 3.5 inches.
(Image by Jeff McLane, courtesy of
Mark A. Rodriguez)

Inspired, Rodriguez decided to seek out tapes of every show the
Grateful Dead played from their inception in 1965 to Jerry Garcia's
death in 1995. As he scanned covers and posted them on his blog, *Dead
Tape Collector*, unsolicited offers of collections rolled in. Some older
batches had fastidiously notated covers all in the exact same format,
while more recent ones were more graphically oriented, with decoration as important as information. "There's this history that I never
really knew was so vast," he says. "It's a history of communal exchange.
The Grateful Dead world has set up some protocols for it, and on a
sociological level, there's a lot of unsaid or learned behavior that comes

with participating in that subculture. Everyone who has tapes is privy to that and has learned that over the years."

Those ideas moved Rodriguez to turn his collection into an art piece that he has displayed at numerous galleries. "It's about playing with value, whether it be the value of the music, the value of the history, the value of the tape as an object," he explains. "It just keeps moving along with these different values inherent to it, and that's just really interesting and fun to pursue, even if it drives me crazy a lot of the time." Part of that continual motion comes from the fact that Rodriguez still hasn't acquired a complete collection of Dead shows. He does have over 3,000 tapes, and he has sold five iterations of the art piece, each time copying tapes from the previous version and reproducing their covers by hand, all while seeking out more. "Every successive generation that I make, I add a few more tapes," he says. "So I'm filling in those gaps, but you're not getting a complete history, and there might be misnotated history on there. And then on top of that, since I'm dubbing tapes from before, the sound itself is changing."

Rodriguez's art project touches on the value people place in music—in who created it, in the history behind it, and in the way we pass it from person to person and generation to generation. It's also a testament to how the cassette tape, as a personal, handmade document, can connect people. "Ultimately, to me, it's a celebration of communication," Rodriguez says. "It's just fascinating that you have this wealth of information from this one highly documented historical entity—and it's all on cassette tape."

CHAPTER FIVE

The Tape Hunters

Traveling the Globe to Unearth History on Cassette

In 1997, Mark Gergis landed in Syria in search of cassette tapes. "The taxi driver dropped me in the center of town at eleven at night," he remembers. "There were so many cassette stalls and kiosks and guys with carts full of tapes walking around. They were blaring their tapes as loud as they could, all competing right next to each other. It was cacophonous and deafening, but it was so exciting."

Though he was only in his twenties, Gergis had long been a music hunter. He grew up in the 1970s in California, where his father, an immigrant from Baghdad, played him Iraqi and Arabic songs. He heard more of the same at family weddings and parties. Once his parents bought him a turntable, he started collecting all types of music. Visiting relatives in Detroit, Gergis discovered imported releases at Iraqi shops. In the early 1990s, moving to the Bay Area and finding its substantial Southeast

Asian population, he dug through markets and grocery stores for sounds from Cambodia, Vietnam, and Thailand.

Though these small, independent shops carried CDs, their cassettes were usually cheaper and more interesting. They were often filled with music from lesser-known genres and adorned with colorful covers featuring artist portraits collaged into vivid or dreamy backgrounds. Store owners often assumed Gergis wanted English-language music or popular Arabic tapes, but he would surprise them by grabbing the most obscure-looking titles he could find. "They were selling to themselves, to their own communities," he explains. "Very few outsiders were coming in and showing an interest in this music. With Cambodian and Vietnamese stuff, I would get the most quizzical looks. They'd say, 'Why do you want this if you can't understand the language?'"

Gergis saved money to travel outside of America but was hesitant to visit Iraq for fear of being conscripted into the military due to his lineage. He instead chose a country nearby, Syria, which he heard was one of the biggest cassette producers in the region. His first impression confirmed that rumor. Armed with a shortwave radio and tape recorder, he immediately started capturing sounds from the airwaves and rifling through kiosk racks, finding everything from mainstream pop to regional Arabic tunes to imported music from Western rock stars. Tapes that bore signs of neglect had the most impact. "I was interested in the dirtier stuff, the stuff that big-city people rejected," he says. "I wanted the tapes you could find on the bottom shelf, the dusty, rat-shit area of the cassette stall." Such finds were plentiful in Syria, and the selection was as diverse as the many parts of the country Gergis visited. At each stop, he would find tapes he hadn't seen in the previous town, by local artists who were known in that village or region but whose work hadn't traveled much farther.

Particularly exciting to Gergis were tapes of wedding music. They documented sounds that likely would never have been preserved were it not for the cassette format—sounds that Gergis calls "anomalous and raw and urgent." Many of these tapes were actually captured live at weddings. But on others, artists recorded wedding-appropriate songs in a studio or in their homes, adding contact info on the cover to create the cassette equivalent of business cards that they could pass around to get booked at more ceremonies. One wedding singer, Omar Souleyman, performed a hard, fast version of dabke, a Levantine form of dance music. Though Souleyman had only been at it for about three

A cassette kiosk in Damascus, Syria,
2006. (Photo by Mark Gergis)

years when Gergis first found his work, he estimates that by then the
singer had already released over fifty tapes. He even had a regional hit
song called "Jani" that mixed Arabic and Kurdish traditional melodies.
"It sounds like, wow, the guy's super prolific," he says. "But actually he's
just working a lot. He's playing a lot of weddings and they're all getting
recorded and released on cassette."

Gergis's initial experience in Syria was so successful that it led to
decades of dedicated international cassette hunting. In this ongoing
pursuit, he often feels like the ultimate find is just around the corner,
just one dirty, unloved cassette away. He will dive into one genre and
find layer after layer, realizing he could spend the rest of his life learning
about that single style. The hardest searches come when he knows that
what he's looking for only existed briefly and probably hasn't turned up
in a shop for over twenty-five years. Titles like that can stay on his list for
decades, and even finding the smallest clues can bring him indescribable
excitement.

"Sometimes collecting is disease and debauchery, and sometimes it's
the elixir for mediocrity," Gergis told Radio Web MACBA in 2011. "Sometimes it's trying to capture a moment in time that can't be reclaimed,
or trying to isolate something out of reach so you can examine it up

The tireless international tape hunter
Mark Gergis. (Photo by Ayesha Keshani)

close, or maybe it's related to an important personal memory or feeling that wants to be relived somehow. We have to stop and ask ourselves why we collect things sometimes. Is it by habit? Is it by some strange obligation? Is it a service or a disservice to ourselves or to others? At its most basic and savage, it's about thrill seeking or competition—and at its most refined, I think it's about self-education and hopefully results in a certain enlightenment."

For Gergis, enlightenment means unearthing music that is at risk of extinction, down to finding the tiniest details of the fastest-disappearing subgenres. It's a goal that for him clearly rewards the amount of time and effort that his global digging requires. "It can be really tough," he says. "You're restricted by space—how are you going to take all this back home? Maybe you're going to three other countries before you leave. Maybe you're not going home for a while. Or maybe you're restricted by time, and you can't listen to everything, so you have to judge just by tape covers, to guess the time period and the genre based on what they look like. Sometimes there's such scant information, not even a label on the tape. Sometimes they're handwritten; sometimes they have some really tactile collage work with cutout photos and text. You learn as you go."

"Sometimes, you come to a small town and you walk around all day in the heat and you have no luck," he continues. "Then you walk by a shop that's run by two people who are each about eighty years old, and they're selling car speakers and stereo equipment and stuff like that. You see remnants of tape racks, but they're full of audio cables or souvenirs. But then you see that behind the counter there's a box with a cassette sticking out of it. And then you start asking questions." Recalling a bountiful trip to a stall in Thailand, Gergis practically beams. "My God, we hit the mother lode on that one," he says. "I think I spent three or four hours in this shop, and I couldn't believe what I found. Doubles and triples of some titles, sealed copies, all kinds of treasures. There's nothing like that feeling."

The feeling Gergis describes is made possible by the cassette tape's remarkable proliferation throughout the world in the decades following its mid-1960s debut. As huge as the format's impact was in the West, it was possibly even greater elsewhere. In many countries, before the cassette came along, the music industry was often so dominated by a few large corporations, and so subject to state control, that independent record labels and distributors were either scarce or nonexistent. Opportunities for artists and producers to spread their music without using conventional channels or expensive promotion were rare.

The cassette tape changed all that. The inexpensive nature of tapes, recording equipment, dubbing decks, and everything else associated with the format meant that almost anyone could produce and release music. All you needed were two recorders (one for recording, another for copying), a microphone, and perhaps a little cash to pay musicians. Tapes could circulate quickly without advertising or airplay, via word of mouth and loudspeakers blaring from kiosks, boom boxes, and taxis cruising the streets. This informal, nearly barrier-less network allowed even the smallest music makers to move enough product to stay afloat or even do well. In some places, simply selling 100 copies of one cassette title was enough to turn a profit, due to the minimal costs involved.

This wave of independent cassette production sparked a profound shift in many regions' music industries. In India, before cassettes came along, 90 percent of music sold came from the near monopoly known as His Master's Voice. But by the mid-1980s, His Master's Voice's share of the market had shrunk to just 15 percent, and most sales came from cassettes on over 300 different labels. In Egypt, an industry previously dominated by state-owned company Sono Cairo saw its total of independent cassette companies double from twenty to forty in just one month. Within ten years at least 500 such ventures were releasing thousands of titles and selling millions of copies. In Indonesia, the national label Lokananta pressed nearly 42,000 vinyl records in 1970; five years later, it sold less than 300 pieces from that stock—and almost 900,000 cassettes.

All of these changes occurred alongside an even greater groundswell of pirate-style cassette production. This included bootlegging of live performances and unauthorized duplication of commercial recordings, of both Indigenous music and records imported from the West. Copyright laws in many countries were either outdated, confusing, unenforced, or simply nonexistent. Anyone willing to invest a relatively nominal amount of money into a rack of chained cassette decks could churn out hundreds of tapes per day and sell them on the street almost immediately.

As a result, in some areas, the majority of cassette sales were of counterfeit tapes. One 1979 study done in France claimed 850 million unauthorized tapes had proliferated around the globe, including 300,000 from Egypt alone; in 1981, another study claimed the worldwide number had increased to 6 billion. Under the direction of President Anwar Sadat, Egypt established a special police force dedicated solely to stopping unauthorized cassette production. In the 1990s, they boasted of confiscating 600,000 tapes a year. In a move similar to the British "Home Taping Is Killing Music" campaign, some Egyptian cassette companies printed text on tape covers encouraging customers to report bootlegging, with phone numbers to call and sometimes a cash reward offered in return for information.

In most of these countries, legitimate labels and bootleggers both thrived by satisfying a market that had rarely been acknowledged before cassettes came along. Styles and forms of music previously ignored by big companies—which could mean they hadn't been documented at all—found a home among independent cassette producers, who could cater more quickly and specifically to local audiences via tapes. Eventually novel genres, born through cross-pollination of these newly exposed forms, emerged on tape. These styles became even more hybridized due to the influence of Western music that made its way into the racks of small-town shops via pirated cassettes.

Take Yess Records, an Indonesian label that released over 700 bootleg tapes from the mid-1970s until the late '80s. Nearly all of its releases were unauthorized copies of copyrighted recordings from other countries, dubbed onto blank cassettes and packaged in a distinctive ocean-blue cover with the label name in bubble letters at the top. The bulk of these were by Western progressive-rock bands such as King Crimson, Marillion, and the band that gave the label its name, Yes. "I really want to make [the label] all about progressive music," said one of Yess's owners, Ian Arliandy. "My mission was to spread this kind of music." The Indonesian government wasn't keen on the country's own music being bootlegged, but it was less concerned about Western releases. So Yess thrived for thirteen years, disseminating music that probably would have missed the country otherwise and likely had an influence on musicians who heard it.

Throughout the cassette explosion, new generations of musicians built followings, and even became stars, by operating outside of mainstream channels, releasing their work almost exclusively on cassette. Music previously scorned by elites, deprived of radio play and shunned from mainstream public performance, now had an outlet. "The cassette revolution had definitively ended the unchallenged hegemony . . . of the corporate music industry," wrote Peter Manuel in *Cassette Culture*, his book on the tape industry in North India. "Most of the new cassette-based musics are aimed at a bewildering variety of specific target audiences, in terms of class, age, gender, ethnicity, region and, in some cases, even occupation. . . . The average, non-elite Indian is now, as never before, offered the voices of his own community as mass-mediated alternatives to His Master's Voice."

In India, these voices came in a stunning variety of forms, revitalizing genres that could have faded out. This included ghazal, a classical style that originally only existed in upper classes but mutated into a broader form with the advent of cassettes; bhajan, a type of ancient Hindu devotional song; and katha, a form of musical storytelling. In areas such as Garhwal, Haryana, and Braj—all not far from the capital city of New Delhi—releases of local music in different dialects came out monthly. According to Manuel, by the 1990s, regional folk styles occupied nearly half of all the cassettes sold in North India. Similar stories emerged around the globe throughout the 1980s and '90s. In the Andes, chicha, which fused traditional huayno music to Western surf and psychedelic rock, flourished on tape. In Iraq, cassettes spread the urban sound of basta, the improvisational vocal form mawwal, and most prominently, the folk dance of choubi. The latter is a variation on dabke that features a rapid, drilling beat, often performed in Baghdad nightclubs. Before cassettes existed, dabke was rarely even recorded, but suddenly thousands of tapes flooded the market.

In Sri Lanka, a dance music called baila, centered around a body-moving rhythm, was boosted by cassettes 400 years after it was first introduced by the Portuguese. "Most Baila cassettes are not deemed suitable for radio performance," explained Roger Wallis and Krister Malm in their book *Big Sounds from Small Peoples*. "But [they] do well in the market even without a large media backup." That success was due in part to taxi drivers playing baila tapes for their passengers, akin to the way hip-hop made its way around New York when blasted from car-service speakers. Local and international music in Sri Lanka also spread in the mid-1980s via thousands of "cassette bars" strewn throughout cities. These businesses sold customers personalized mixtapes with songs they selected from an in-house catalog.

In Indonesia, music from the Banyumas region of Central Java gained new status on cassette, including rural genres such as EBEG (a.k.a. "hobby-horse trance dance") that were previously considered unworthy of attention. "Now that cassette companies find that Banyumas music will turn a profit for them, Banyumas tradition is gaining a measure of prestige and recognition that would have been impossible during earlier times," wrote R. Anderson Sutton in a survey of Javanese cassettes. "It enjoys more nearly the same system of patronage as the mainstream tradition, packaged in the same colorful little cassette cases and sitting on the same shelves in the cassette stores. . . . In these cases, the cassette industry acts as a leveler, blurring the older status distinctions that were still in place a generation ago." Tapes didn't just spread music like this; they broke through caste barriers, overriding previous rejection of working-class performers by well-off elites.

Along with helping to blur societal boundaries, cassettes also served as rebellion against the standards of what governments deemed acceptable. In Israel, music created by Mizrahi Jews, some of which incorporated styles from Europe and North Africa, reached listeners via locally produced cassettes, because state-owned radio generally avoided it. In Turkey, a working-class genre called arabesk that blended Indian and Arabic influences was banned from television and radio, but it circulated on tape, often heard on buses driving between towns. And in many countries, protest music that targeted the restrictions and oppressions of the government spread primarily through cassettes. Take the dissident rock of Chinese singer-songwriter Cui Jian, whose music influenced students during the 1989 Tiananmen Square protests. Or the Cuban leftist folk singer Silvio Rodríguez, whose tapes were snuck into Chile during the reign of dictator Augusto Pinochet.

Perhaps the most fascinating cassette-fueled battle between music and the government happened in Egypt in the 1970s and '80s. As tapes of local and regional music spread—particularly a working-class style called shaabi, meaning "of the people"—the state and the press reacted with hyperbolic horror. Journalists deemed the musicians "art imposter clowns" and the music "incompatible with public taste," even claiming that it was more dangerous than cocaine. "Audiocassettes, in the opinion of many critics, facilitated the spread of 'vulgar' sounds by making it possible for anyone to be an 'artist' regardless of his or her training," explained Andrew Simon in *Media of the Masses*, his book on Egyptian cassette culture. "In enabling any citizen to become a cultural producer, as opposed to a mere cultural consumer, cassettes, they claimed, lowered artistic standards and tarnished public taste."

Despite this propaganda, or perhaps even because of it, cassettes boomed on the streets of Cairo. People with non-music-industry jobs— carpenters, electricians, cab drivers—started their own labels and churned out tapes. "Everyone who enjoys his voice sings, and issues a collection of new songs every two months," an Arab singer told the weekly magazine *Ruz Al-Yūsuf*. One of the earliest "vulgar" musicians, Ahmad Adawiya, became successful so fast that his first tape, 1973's *Al-Sah al-Dah Ambu*, sold a million copies. He became the face of the cassette-based music that swept across Egypt, to the chagrin of cultural gatekeepers.

Besides being cheap to make and promote, cassettes proved to be the only conduit via which so-called vulgar music could make it to Egyptian ears. State-owned radio exerted strict restrictions on what made it to air, with two screening committees in place to keep listeners away from the supposed dangers of independent music. In the government's Office of Art Censorship, lyrics had to be submitted for approval and later checked against a final recording. But the department only had fifteen censors and seven tape decks with which to screen the overwhelming amount of music produced in Egypt and coming in through its borders. Even specialized national councils created to determine the "purity" of tapes and declaring "war on the cassette" couldn't stem the tide. Cassettes continued to blare music from sidewalks, cars, buses, and shops, playing everything from the smallest regional genres to the biggest mainstream acts.

"By offering any citizen a means to record their voice and to reach a mass audience, cassette technology enabled an unprecedented number of people to create Egyptian culture at a time when public figures strove to dictate the shape it assumed," wrote Simon. Around the globe, the cassette tape facilitated a new democratization of the music industry, allowing artists of any age, gender, education, or class to bypass corporate channels and state control, reaching the ears of people directly.

But how can non-Western cassettes reach Western ears? To hear these tapes outside of the places where they were made, do you have to be a dedicated traveler and digger like Mark Gergis? It's a question he has often pondered himself, especially when he began connecting with other tape hunters in the early 2000s. "We all came together with our music and our stories and said, 'What the hell?'" Gergis recalls. "'Why don't we find any of this music anywhere in the West? Why isn't it available?'"

Of course, some international music has reached the West in official, sanctioned forms. But that's not usually the kind of music Gergis is searching for, since well-funded releases are often filtered by state officials on one end and academics on the other. "It's pretty sterilized," explains Gergis. "When an American label or organization goes to a foreign government and says, 'Give us your best music and we'll put it out,' what do they want to be known for? Not in a million years would it be anything on these cassettes. That's not the image that they want to project. So the 'world music' that's sold in the West . . . it's like getting Indian food outside of India, there's just no kick to it. You're up against this cultural delegation, from both sides."

Gergis found a chance to help correct that situation when he met Alan Bishop, who has been traveling the globe in search of unheard sounds for even longer than Gergis. Bishop grew up in Michigan in the 1970s in a family with Lebanese ancestry. His grandfather would often play traditional Arabic music for him and his brother Richard. The siblings moved to Arizona in the early 1980s, forming an avant-garde rock trio called Sun City Girls. Not long after, they traveled to Morocco, where Alan stayed for two months, spending most of his time collaborating with local musicians and recording radio broadcasts onto cassette tape. "Cassettes were the main source of music for sale in every shop, everywhere I would go," Bishop remembers. "I would turn on the radio

Alan Bishop, musician, documentarian,
label owner, and relentless world traveler.
(Photo by Hans van der Linden)

and start recording snippets of music I was interested in, and then take
that into a cassette shop and ask the clerk what it was. Often they would
know right away and have a cassette of it there to sell me."

Soon, Bishop started visiting other countries in search of tapes.
"Everywhere I went, I found streets where all I would see was row after
row of cassette stands," he remembers. "Bangkok in Thailand, Jakarta
in Indonesia, Aswan in Egypt, they were all like that." One of Bishop's
most eye-opening moments happened in the latter city, which sits in
the south of the country next to the Nile River. During a 1985 visit, he
suddenly fell ill. Heading to his hotel to rest, he noticed a huge cassette
stand and couldn't resist. "I was about ready to pass out, and I just said
to the clerk, 'Choose two or three cassettes for me and I will buy them,'"
he recalls. "He sold me three tapes, I went back and played them in my
room, and they were completely amazing. The guy just knew what I
wanted. I don't know how, he just knew." Most of the time, though, tape
hunting isn't that easy. "You're not going to find much unless you work
for it," Bishop says. "It's a job, and it's a passion. You have to be willing to
love what you're doing." It shows: Bishop has traveled so much that he
claims, for example, to have been to Thailand over thirty times.

The band Bishop started before he launched into most of his hunting—Sun City Girls, with brother Richard and drummer Charles Gocher—eventually began drawing from his international discoveries. They mixed covers of the songs on Bishop's tapes into their whirling repertoire of rock, folk, noise, and improvisation. Taking another cue from what Bishop had found, most of Sun City Girls' early music was recorded at home on boom boxes, handheld recorders, and 4-Tracks and released by the band themselves on cassette. "We had so much stuff and I just wanted to get it out there," Bishop recalls. "Cassettes give you the power to control your music—what you put out, the artwork, the price. You can spend $100 to make some, and if you make a profit you can make some more. You can do more extreme stuff that no label would put out, the most fucked-up shit that you could ever imagine."

While Sun City Girls' music was spreading, Bishop was also trading dubs of the music he found overseas with a group of fellow hunters that included brother Richard, Mark Gergis, North African–born documentarian Hisham Mayet, and Seattle-based ethnographer Robert Millis. In 2003, Bishop and Mayet started the Sublime Frequencies label, inviting contributions from the rest of their circle. Initial releases included compilations of music from Sumatra, Java, Bali, and Burma, as well as releases taken from Bishop's cassette mixes of radio broadcasts. The sounds on the label were diverse from the start: their first compilation, *Folk and Pop Sounds of Sumatra, Vol. 1*, featured songs in five different regional styles.

In 2004, Gergis made his first contribution to Sublime Frequencies, a double CD called *I Remember Syria*. It was motivated by his desire to counter the American view of Syria as a rogue terrorist state and instead celebrate the country's vibrant cultural history. "When the U.S. began ramping up its rhetoric against Iraq in 2003, I was angered, and sensed Syria might be next," he told the *Quietus* in 2013. "This pushed me to assemble *I Remember Syria* with greater urgency. I like to say that it was an audio love letter to the Syria I grew to know." The album is a collage of material Gergis collected in his first two visits, including sounds from street performances, weddings, mosques, interviews, and radio broadcasts.

Later that year, Gergis compiled another Sublime Frequencies release, *Cambodian Cassette Archives: Khmer Folk and Pop Music Vol. 1*. This time, the music was not culled from his travels but from over 150 tapes he found in an Asian branch of the Oakland Public Library. When Gergis brought the completed album to one of his favorite Cambodian cassette stalls, the owner was shocked. He himself had distributed nearly all the artists on the compilation and couldn't believe there might still be any interest in them. "He put all of these recordings into the garbage bin years ago," Gergis told Radio Web MACBA. "Which is where, in his mind, they belong."

That story points to an interesting aspect of Gergis's and Bishop's efforts. Through Sublime Frequencies, they seek to resurrect cassette music that was often treated as disposable in its own time and place, discarded to make way for whatever the next hot tape might be. "The regional tapes were ephemeral as hell," says Gergis. "Often the shops would write the year on the cassette, so it would say 1998 in bold letters on the front, and no one would want to buy those in 1999. If I went back a year later, those were gone, either sold or thrown out. So you have a real problem in terms of saving this music, because it really did live and die on cassette." Tape artist Aaron Dilloway confirmed this problem while living in Nepal in 2022. When visiting the country twenty years earlier, he saw cassette shops everywhere. "Now there are barely any," he laments. "I went to a place still called Bouda Cassette Center, but they had about ten cassette titles buried in the back. I asked the woman there, 'Where did they all go?' She said, 'We threw them all away.'"

"It's a matter of economy and standard of living—it dictates how much of a cultural legacy that can be not just preserved but promoted," Bishop told the *Believer* in 2008. "The culture is left to rot, just like the buildings and the infrastructure. No money is going back into preserving things."

Over the course of their tape hunting, Bishop and Gergis discovered some artists they found so compelling that they were inspired to track those individuals down. In the 1980s, Bishop found a tape of Egyptian instrumentals and played it often in Sun City Girls' van during their tours. Over a decade later, he figured out the music was by Omar Khorshid, an Egyptian composer and guitar player who made albums in Lebanon during the 1970s before dying in a car crash in the early 1980s at age thirty-six. Sublime Frequencies has since issued two Khorshid albums, including one recorded just days before his passing. Richard Bishop also wrote and recorded an album, *The Freak of Araby*, that includes multiple covers of Khorshid songs.

Another artist whom Gergis was fascinated by turned out to be still making music. In 2006, about a decade after first discovering tapes by wedding musician Omar Souleyman, Gergis went searching for the man himself. Through a chain of phone calls to people who know people who know Souleyman, Gergis found him and convinced him to release music on Sublime Frequencies. Within a few years, Souleyman began touring the West, and his music reached people who would never have been exposed to his sounds had they been left to gather dust at the bottom of Syrian cassette racks. "Omar unwittingly became the first-ever Syrian singer to 'break' on the scale he did," Gergis told the *Quietus* in 2013. "And for the first time, it wasn't the result of flash producers trying to add global beats and fusion into the equation. This was the real deal from rural Hassake, Syria—and it had a power of its own that didn't need any of that."

On a trip to Syria in 2010, Gergis, who at that point was Souleyman's manager, still kept an eye out for Souleyman tapes. At one stall, he inquired if the owner had any Souleyman cassettes and was surprised by the tape that turned up. "The front cover showed Omar in front of the Eiffel Tower, photographed by me," Gergis recalls. "Here was a photo I had taken of a musician whom I had discovered through tapes, and now my photo was on one of his tapes that I discovered. So it all just came full circle."

Over the years, Gergis's collection of Syrian tapes has stretched to over 400 titles. In 2019, he began to digitize and upload them to his website, Syrian Cassette Archives. It's a project about preservation, as well as a reminder that some of the cultural history of many countries can be found only on the cassette tapes that made their preservation possible in the first place. "It matters because the tapes may tell a story in the future that needs to be told, to Syrians or to the wider world," Gergis told *Financial Times* at the time of his website's launch. "It becomes part of a cultural collective memory that might fade away otherwise."

Gergis and Bishop hunt for vinyl too, and most Sublime Frequencies releases come out on that format and CD. But in excavating musical history on cassette, they push back on the conventional narrative that the most exciting finds can come only on vinyl. Gergis recalls a visit to Japan in 2013, years after Sublime Frequencies had issued compilations of molam music, a psychedelic take on a Thai rural folk form. "These kids were asking me, 'Where did you find all those records? We've bought every record you could buy in Thailand!'" Gergis recalls. "I told them

Some cassettes Mark Gergis has found
in dusty stalls and kiosks around Syria.
(Photo by Mark Gergis)

those weren't on records, they were on cassette. And they all looked at each other like, 'Oh my God, we've been barking up the wrong tree!' There's a lot of music that only went to cassette because that was the only affordable option. You literally have a whole era that's a cassette era. That's where that music lives."

"I run across a lot of vinyl collectors and they're so proud of how they dig out the greatest vinyl from Thailand or Indonesia, and then they say, 'Oh, I don't mess with cassettes,'" adds Bishop. "I just laugh and say, 'Then you're cutting yourself out of an amazing amount of incredible music in the thousands of releases that were put out on cassette only between 1970 and 1995. You're just gonna miss all that music? You think you have it all? You're wrong.'"

"I think Africa is even more diverse than most people realize," says Brian Shimkovitz. "The cassette really made it possible for diverse African music to come out and possible for people who traveled to hear the music." Shimkovitz knows this from his own journeys. In the early 2000s, his studies in ethnomusicology at Indiana University took him

on a trip to the West African country of Ghana. There he found numerous vendors selling cassettes, a format he developed an affinity for when he began collecting Grateful Dead live recordings as a teenager. Shimkovitz quickly began snapping up tapes, most of them priced at a dollar apiece. "I wasn't really thinking of it as collecting at the time," he says. "I was trying to get lots of different stuff, to get an idea of what was out there, even if I didn't have time to listen to it immediately. Not just the most obscure music, but also what I was hearing on radio and television. And one of the greatest things about shopping for tapes in West Africa is that they always would let me listen to them first. Even if it was sealed, they would cut it open and play it on a boom box."

It turned out that getting a wide range of music wasn't that hard. Shimkovitz found everything from unlabeled tapes and deceptively authentic-looking bootlegs to cassettes bearing music-industry hologram stickers as proof that they were legitimate. Sending several shoeboxes full of cassettes back home, Shimkovitz became hooked and continued to dig for more on subsequent trips to Ghana, one of which lasted a year. Throughout his visits, Shimkovitz noticed that selections would vary greatly from town to town and store to store, reflecting the wide range of culture in Africa. "You have these countries that were created by colonial powers that don't make any sense in terms of the nation-state borders, so they have hundreds of languages," Shimkovitz explains. "Even a country such as Ghana, which is the size of Oregon or Pennsylvania, when you go to the north, you're in a zone where you're hearing music in totally different languages on cassettes that you can't find in the south, because they're in styles that aren't popular there. Some of these language groups don't get play in the state media, or in the commercial media based in the capitals. But if people from that region speaking that language live in a capital city, they want to hear their music."

By 2006, having moved to Brooklyn, Shimkovitz decided to start sharing his collection. He started a blog called *Awesome Tapes from Africa*, posting digital files of music to download, pictures of the tapes the music came from, and descriptions of what you would hear if you grabbed the MP3s. Just clicking through the site's archives of tape covers reveals a fascinating array of names, titles, colors, and instrumentations. Alhaji Sir Waziri Oshomah poses in a velour double-breasted sports coat, smiling above yellow and red letters that spell "Happy Christmas." There's a man named Mr. Guitar sitting in a baby blue suit, surrounded by a foggy halo. The singer Iliss Ntmazirte is resplendent in an orange shirt with purple flowers in front of a red floral arrangement. Morocco's El Miloudia beams behind her gray-and-white violin while decked out in a sweater and blouse to match. Surrounded by earth-toned abstract art, Sudanese singer Mahmoud Abdel Aziz stares from under slick sunglasses, wearing headphones and a yellow-and-orange plaid shirt. Everything on Shimkovitz's site is so eye-catching and intriguing, it feels like you can already hear the music before you've even downloaded the files.

As word of Shimkovitz's blog spread, people began sending him tapes, and his collection grew quickly. He guesses that by the early 2000s he owned at least 5,000 tapes from Ghana and other parts of North Africa. Eventually, he decided that some of his finds deserved more permanent documentation than a blog could provide. So in 2011 he started a record label also called Awesome Tapes from Africa and began seeking artists' approval to reissue their cassettes. To track them down, Shimkovitz often blindly digs around, cold-calling stores to ask if anyone has heard of the person he's looking for. Even if he finds them, widespread piracy means that figuring out the source of the music can be a hurdle too. "I've had situations where I bought something on cassette and managed to find the artist, and they said, 'How can that even be my music? I never put my music on cassette,'" says Shimkovitz.

In some fortunate cases, the process has been easy. In 2013, Shimkovitz was browsing a shop in Bahir Dar, Ethiopia, when he came across a 1985 cassette by Ethiopian musician Hailu Mergia called *Hailu Mergia and His Classical Instrument*. The shop clerk let him preview the tape, and Shimkovitz was immediately entranced by the way Mergia weaved together accordion, Moog synthesizers, and a drum machine. He plugged Mergia's name into a web search engine, found a phone number, called it, and Mergia answered. It turned out he had been a regular on the music circuit in Addis Ababa, first as a member of a group called the Walias Band. But in the early 1980s, he had moved to America and become a cab driver in Washington, DC. There, he recorded *Hailu Mergia and His Classical Instrument* by himself and sent it back home, where the cassette became a surprise hit. He was happy to let Shimkovitz release it again, and Awesome Tapes has since reissued more titles as well as new music by Mergia, who has used the renewed interest as impetus to begin performing again.

In other cases, Shimkovitz has become a detective, chasing leads, finding translators, and working to make sure artists approve and get paid when their reissues sell. In his very first blog post, he wrote, "You may never hear anything like this elsewhere. I bought this on the street from a guy selling tapes displayed on one of those big, vertical wooden racks in Cape Coast, Ghana." The song was by an artist named Ata Kak, sung in a language known as Twi, and played in a style called hiplife. Hiplife mixed rap and a Ghana-based form called highlife, which Shimkovitz happened to be studying. The track came from a 1994 tape called *Obaa Sima* that Shimkovitz had found in Ghana in 2002. But info on Ata Kak was scarce: his real name and contact info weren't anywhere on the tape, and nobody in Ghana whom Shimkovitz talked to had ever heard of him.

Shimkovitz's search for Ata Kak took over a decade and was so full of twists and turns that a short documentary was made about it. He traced leads from Los Angeles to Dusseldorf and Canada, once blindly flying to Toronto when he heard that Ata Kak had lived there when he made *Obaa Sima*, which was originally released in an edition of just fifty copies. Eventually, the discovery of Ata Kak's son on Facebook led Shimkovitz to the musician, whose real name is Yaw Atta-Owusu, living in Ghana. The tape was officially reissued in 2015, sourced not from the original master, which was too degraded to use, but from the cassette Shimkovitz himself had found. "It was just a lonely cassette at a market stall," wrote Charlie Frame in a review of the reissue for the *Quietus*. "But it beggars [*sic*] the question: Who exactly is going to trawl through the rough of thousands upon thousands of demos, false starts and plain old trash found online until the next diamond in the rough is found?"

Shimkovitz himself has no plans to stop being an answer to that question. He continues to search the globe for tapes even though they've become harder to find as digital media has taken over. "The last time I was in Ghana, there was this girl in the market who was selling DVDs, and I was like, 'Where are the tapes?'" he recalls. "And she said, 'Oh my dad already put them back in our family warehouse.' They'd been mothballed." Refusing to do the same with his own vaults, Shimkovitz DJs with tapes in dance clubs and other performance spaces. Using two tape decks with pitch controls, he melds the disparate styles across his collection into a danceable mix. "It's distinctive to be playing tapes, but I think it's more distinctive to be playing this type of music," he says. "What I really appreciate about the cassette tape is how many different voices it has helped to get out there and be heard."

Shimkovitz has found cassettes from all over Africa and across the globe, but his focus has been in North Africa, where there is certainly more than enough material to continue searching. Chris Kirkley has found the same to be true of West Africa, particularly a region called the Sahel, which covers over a million square miles and includes parts of Senegal, Mali, and Sudan. Kirkley first traveled there in 2010, with the simple goal of recording music with a portable tape deck just to learn more about the area. He ended up staying in Mali for a year and a half, and returning there from his home in Portland annually after that. "There are very strong traditions for localized music in this region, mainly because there are so many different languages and ethnic minorities," he explains. "There are eighty languages in Mali, and people forge their identities based around their language groups."

Kirkley became particularly enamored of the music of the Tuareg, a somewhat nomadic group whose history dates back thousands of years. In recent decades, a version of Tuareg music driven by guitar and influenced by Western rock music emerged. "It was really created because of the cassette tape," Kirkley says. "A lot of it came from Libyan rebel camps with people recording music on cassette, redubbing it, and trading it throughout the diaspora. It could get you a jail sentence if they caught you with that kind of cassette. So the cassette format was really important."

At first Kirkley just wanted to share the music he had recorded with friends back home. But Portland store Mississippi Records got ahold of one of the homemade compilations he'd been passing around—culled from three regions of Senegal and Mali—and inquired about releasing it. Kirkley agreed, launching his own Sahel Sounds label to partner with Mississippi. In 2010, they turned his compilation into his imprint's first release, *Ishilan N-Tenere: Guitar Music from the Western Sahel*. Since then, Sahel Sounds has released material from numerous Tuareg artists, both archival recordings and new works, and brought some to the United States to tour, most notably the guitarist Mdou Moctar and the female-fronted group Les Filles de Illighadad.

Some of the Tuareg artists Kirkley has found were themselves influenced by cassettes, in the form of bootleg releases from Western rock bands. "Dire Straits [the British rock band famous for Mark Knopfler's circling guitar lines in the hit "Sultans of Swing"] in particular is huge for Tuareg guitarists," he says. "Almost every Tuareg musician cites them, and I've actually seen their tapes everywhere." Another Western artist, Prince, had enough impact on the region that Kirkley himself directed a new version of that musician's movie *Purple Rain*. Filmed in Niger and starring Mdou Moctar in the role Prince once played, the film is called *Akounak Tedalat Taha Tazoughai*, which translates as "Rain the Color of Blue with a Little Red in It" (there is no word for purple in Air Tamajeq, one of the Tuareg languages).

One of Kirkley's most fascinating discoveries is Mamman Sani Abdullaye, a Nigerian innovator in the realm of electronic music. Inspired by folkloric Nigerian songs, Abdullaye made an album in 1978 called *La Musique Electronique du Niger*, filled with hypnotic organ instrumentals that seem to emanate from another planet. "It had this incredible picture of a guy playing an organ, and the music blew my mind," says Kirkley of the copy he originally found and reissued in 2013. "I was able to track him down the next day and found out there were, like, two copies of this tape still in existence. I don't even know how many they ever made—probably 100 at most." Such scarcity makes much of the music Kirkley seeks a kind of cultural endangered species, providing continual motivation as Sahel Sounds' catalog stretches past fifty releases. "There's an era of music that came out on cassette tapes that could be lost because none of the original tapes survived," he says. "There are almost no master recordings of this stuff. Whatever we can find or rescue is important."

The vast number of cassettes still to be rescued outside of the West seems infinite, but Western countries have their share of hidden gems too. Considering how many individual artists have created homemade tapes since Lou Ottens's invention in the 1960s, in small editions or even just as a single copy, the number of cassettes that could still be out there somewhere is daunting. For Jed Bindeman, it's also thrilling. "I'm obsessed with this idea of finding fully undiscovered music," he says. "There's just so much music out there that came out on cassette specifically that's essentially been lost, that no one's rediscovered yet. It's wild to think about how many people over the years have made weird music—thousands and thousands—but the amount of people who have actually heard it is so few. It's so much fun to dig around for and think, what is this stuff?"

Born in the early 1980s, Bindeman caught the tail end of the underground tape era as a teenager, when it was still possible to grab a copy of a magazine and send some money or postage to an address in the back in hopes of getting back a great tape. As he got older, Bindeman started collecting vinyl, but in the mid-2010s his interest in cassettes was rekindled. "With records, it seems like now most things have been found to some degree," he says. "But with tapes, it's still like the Wild West out there, filled with cassettes that came out in editions of fifty in 1982 or something like that. If you find one of these, it might be the only copy that exists. It's exciting to hold that artifact and think, 'Wow, if I wasn't listening to this thing right now, chances are this would never be heard ever again by anyone!'"

Bindeman has spent nearly a decade trawling record stores and thrift shops, as well as tracking down people who made small-run tapes to see if they have any left. He documents his finds online via an Instagram account under the moniker Concentric Circles, posting pictures of cassettes along with snippets of the music they contain. His collection consists mostly of handmade, independently produced tapes that are filled with personality in both packaging and music, giving a vivid sense of how many people have found the cassette tape format perfectly suited to artistic self-expression. "Cassettes are the most unpretentious format out there, where you can find kind of the most genuine, undistilled music that was ever made," he says. "It's the most utilitarian 'of the people' format that was ever created. It put everything in people's hands and they could just go crazy."

Bindeman also launched a record label called Concentric Circles, inspired by one of his most unlikely finds. One day his girlfriend Natalie Howard was sifting through a bin at a Goodwill secondhand shop near their home in Portland, Oregon. "It's a big warehouse where they have carts full of unsorted stuff," Bindeman explains. "They wheel out a cart for maybe two hours before they throw it all out." Howard noticed a stack of handmade tapes and brought them all home. One near the bottom of the pile, which had handwritten song titles and a photocopied picture of a woman folded inside, caught Bindeman's attention. Its music featured an intense, haunting voice singing emotional lyrics over a backdrop of minimal, desolate synthesizers. "Each song was better than the last," Bindeman recalls. "I became quite obsessed with the tape from that point out and did my best to try and find her."

The only information on the tape was the woman's first name, Carola, and a phone number. That number was long out of service, but Bindeman discovered through Facebook that Carola, whose last name is Baer, was now a teacher living in England. It turned out she had recorded the tape in San Francisco in the early 1990s by herself at home, with no intention to release it. It was actually the only copy she ever made. "She told me that she'd thrown away probably sixty tapes like that over the years," Bindeman explains. "With the songs that were on the tape we found, she figured there were no other surviving copies of that music— no backups of any sort." Had Howard not grabbed the tape, it would have ended up in a landfill, never to be heard again. Instead, Baer's music became the first release on Concentric Circles, under the title *The Story of Valerie*. "It's amazing that it actually fell into the hands of someone who liked it and wanted to do something with it," Bindeman marvels.

Thrills like that are what make cassette hunting such an obsession for Bindeman, giving him a chance to almost literally travel back in time. "It's kind of living vicariously through what these people were doing, piecing it all together and trying to visualize this world," he says. "It makes me think about the way people disseminated their music before the internet was even an option. It was very small and contained; if you sent out a tape to fifty people, you didn't expect much to happen, and probably very little did. Whereas nowadays when somebody just throws up a random song on the internet, anyone might hear it. It was just a different mindset back then. I think people just expected less and were less overwhelmed by the possibilities."

"You have to be patient with tapes, because it takes forever to fast-forward through them to find tracks," he continues. "It's not like a record where you can drop a needle around it quickly. It has made me realize, when I get tapes, even if they look kind of bad, I need to spend time with them and listen to them because my ideas of what things are and what I like changes over time. And sometimes you're surprised by what you find inside."

CHAPTER SIX

The Tape Makers

The Culture of Personal Mixtapes

Menghsin Horng is riding around the suburbs in southwest Michigan, reclining in the passenger seat while her friend Saryah drives. It's the early 1990s, they're teenagers, and there's nowhere they have to be. The soundtrack to their wandering cruise is a cassette rolling along with them in the car's tape deck. Out blasts a string of punk rockers that Horng had read about but never heard before—the Dead Milkmen, Pansy Division, Dead Kennedys' Jello Biafra—plus a few curveballs such as Steve Martin's novelty hit "King Tut." The tape came unsolicited from a kid in California named Jimmy Calloway, with whom Horng had previously traded copies of their respective self-published, hand-stapled fanzines.

Sometime after they've flipped the tape to its second side, the music ends abruptly. Horng figures Calloway ran out of imagination (or records), but she and Saryah have more time to kill, so they keep driving.

"All of a sudden, through the silence of side B, we heard a young male voice call us out for listening to dead air," Horng remembers. "[He was] wondering out loud who was still listening. . . . Maybe I was as bored as he was. . . . Otherwise why let a blank tape continue to roll? Saryah and I both started laughing and spazzing about this weird geeky boy with interesting musical taste who had just embedded himself into our night by total surprise."

Calloway's tape, the first Horng had ever gotten from afar, made such an impact that it launched her into a lifetime of making and trading personalized tapes. "I learned to reach out with music, over and over again, from his first kind gesture," Horng says. "Sending a mix of personally curated songs, even if you couldn't fill up an entire tape, seemed like such a generous gift." In this, she joined an informal club of people around the world recording mixes of music by hand onto blank cassettes, collaging sounds from whatever variety of sources they could find and in whatever order their imaginations conjured. The process of creating idiosyncratic, one-to-one musical missives—known most commonly as mixtapes—became a cultural phenomenon, one accessible to a degree unimaginable before the advent of the cassette. Anyone with a deck, a blank, some music, and a little bit of time could make a mixtape.

Throughout the 1980s and '90s—and beyond, for those truly devoted—people like Horng changed each other's lives with mixtapes. They selected and arranged sounds for friends as gifts, for crushes as courtship, for fellow music fanatics in impassioned exchanges of art and information that spurred learning and discovery. They did this with relatively cheap blank tapes they could buy almost anywhere, not just at music stores but even at groceries and pharmacies. They used everything from the most generic brands such as RadioShack's in-house version up to the two staples of most mixtapers' arsenals, TDK's SA or—if they had a little more to spend—Maxell's XL-II. Meaning and messages could be embedded in every choice: what songs to pick, what to start with, what to end with, how to order and arrange them all, what to write on the cassette, what to put on the cover, even something as seemingly mundane as what length of blank tape to use.

"Were you making a disposable thirty-minute declaration, a heartfelt sixty-minute soliloquy," wrote Christopher Sutton in the cassette-themed issue of the magazine *Basic Paper Airplane*, "or were you possibly hoping to change someone's worldview through 120 minutes of incendiary musical diatribes?" "It takes time and effort to put a mixtape together," wrote musician Dean Wareham in Thurston Moore's compendium *Mix Tape: The Art of Cassette Culture.* "The time spent implies an emotional connection with the recipient. It might be a desire to go to bed, or to share ideas. The message of the tape might be: 'I love you. I think about you all the time, listen to how I feel about you.' Or maybe: 'I love me. I am a tasteful person who listens to tasty things. This tape tells you all about me.'"

As with so many other uses of the compact cassette, mixtapes wrested control of when and how music could be heard away from the companies who produced and marketed it and into the hands of the people who appreciated and loved it. Suddenly, you didn't have to be content just to consume music; you could rearrange it, curate it, mold it to your personality and aesthetic. "I believe that when you're making a mix, you're making history," wrote Rob Sheffield in his memoir *Love Is a Mix Tape.* "You ransack the vaults, you haul off all the junk you can carry, and you rewire all your ill-gotten loot into something new. . . . When you stick a song on a tape, you set it free." Making a mix turned you into a kind of instant DJ for audiences of one, even if that one was just you. This ability to break away from the prepackaged limitations of industry products and the market-wary programming of radio stations had a lasting cultural effect. Mixtape making literally changed the way people interacted with and listened to music.

The concept of making and trading tapes didn't start with the cassette tape format. In the 1950s and '60s, correspondence clubs sprouted up among reel-to-reel enthusiasts. Manufacturers even made short reels of tape specifically for this purpose, with packaging that included spaces for writing addresses and affixing stamps. But for the most part, these postal exchanges involved audio letters, with messages spoken by human voices rather than selections of music, making them a kind of aural version of pen pal correspondence. Once the compact cassette caught on with the general public, the phenomenon of people making musical mixtapes for each other became much more common and much more individualized. It was no longer the realm of audio aficionados who had the budget and expertise to buy and use reel-to-reel players or expensive professional cassette decks. It instead welcomed all kinds of music fans, thrilled with the ability to express themselves to each other through their own ways of selecting, juxtaposing, narrating, and packaging the music they loved.

Mixtape exchanges are usually done from person to person, outside of any spotlight. But some famous musicians are renowned mixtape fanatics. The Clash's Joe Strummer, Sonic Youth's Thurston Moore, and Black Flag's Henry Rollins all accumulated collections that they have shown off in documentaries and even art galleries. In the Beastie Boys' book about their own history, Adam Horovitz opens with an homage to mixtape making. "If the tape you were making was meant to be a gift for someone, it took serious focus," he writes. "It was rarely just about sharing songs via a cassette. It's way bigger than that. Artistic decisions, emotional decisions, and depending on who you're making it for, strategic decisions went into its making. . . . You'd want your tapes to have cool covers like your LPs, flyers, and fanzines. It was a process."

Because mixtapes move one individual at a time, everyone is an equal participant, no matter how well known they might be. Unlike other kinds of tapes that circulate in music scenes, such as demos and DJ mixes, there is no need to know someone on the inside or cultivate a following. All you have to do is make a mixtape, and you are a mixtape maker. Even the most intense, dedicated participants are not famous for making mixtapes, except perhaps among the people they send them to. In a way, mixtapes—like mail art, which they often are a part of—are a form of outsider art, made and exchanged by people who know each other personally rather than publicly.

The one-to-one quality of mixtapes makes them a perfect format for musical self-expression. Music as an extension of personality—either in your record collection or your favorite songs on the radio—wasn't a new concept when the cassette came along. For many, what you listen to is a part of who you are. In this sense, making a mixtape could be an act of personal definition, of marking time, of reflecting and assessing the world like you're writing a diary, even if you're using other people's words and sounds. "I find mixtapes to be more useful little time capsules to remember my life by than whatever weird scribbles would result from each individual day," wrote journalist and mixtape maker Erin Margaret Day on her website. "I spend so much time on each one that I can always return very intensely to the time and space in which I was making that tape, what I was thinking, feeling, processing, considering, or envisioning."

The cassette tape offers a way to tangibly share that personality, to communicate through music, perhaps even to merge friendship and fandom. It fuses the close connections people form to art and to each other into a potentially endless conversation. "There's no format more human than the cassette. No format wears our stain better," wrote critic Nick Sylvester on the music website *Pitchfork*. "When you make a cassette from scratch—the music, the dubbing, the labels, the art, the liners, even the casing—these little human imperfections accumulate in a way that makes the music mean something different." "[A mixtape] does a better job of storing up memories than actual brain tissue can do," wrote Sheffield. "Every mixtape tells a story."

"Songs and bands and musicians are sort of like friends," says Sarah Grady, a devout mixtape maker since her teen years. "You listen to certain musicians to get something out of that relationship that you get differently from another musician. You're relating to them in unique ways that have something to do with who you are as a person. I think friendships are the same, where different people mean something special and unique to you. The mixtape is a combination of those relationships, so they reflect one another."

Mixtapes first entranced Grady when she was growing up in the 1980s in York, Pennsylvania, searching for ways to connect to like-minded people. One of the first solutions emerged when she and her older sister began writing to people who listed their addresses in the back of British teen-targeted music magazine *Smash Hits*. That soon led them to a mail-based exchange known as "friendship books," a homemade combination of pen pal correspondence, classified advertising, and mail art. One person would take a small, blank book and write their address and things about themselves—especially what music they liked—inside. They would then send it off along a chain of people, each of whom would add their own information and pass the book on. "It would get returned back to you after it got filled up," Grady remembers. "You could then look and think, 'Oh, this person on page five really loves Throwing Muses. I love Throwing Muses too!' So you would write them and start to trade tapes with them of music you thought they might like."

Making and trading mixtapes accelerated for Grady when, in ninth grade, she got a dual-cassette deck to dub songs back and forth onto mixes. Every Saturday, she went to the mall to buy one blank tape, using money she saved by skipping lunch at school during the previous week. Eventually, she was trading mixtapes with nearly thirty people, becoming so busy that she kept a calendar detailing whom she owed tapes and when she might get some in return. "I got to spend whole weekends working on one tape and I loved every second of it," she recalls. "You're spending forever trying to pick which of the three songs on a given tape might be the right one to put in front of another song, or you'd get half-way through the first side and think, 'I don't like it, I'm gonna start all over!'" Grady's fellow mixtape traders were similarly devout. One friend in Phoenix, Arizona, shared her love of fastidiously annotating each mix-tape. "For every song, we wrote copious notes about the music and the band, and why we chose that song for that person," Grady remembers.

Such personalized communication permeated every aspect of mix-tapes. Reams of unspoken information could be conveyed in the way a mixtape maker added their own handmade decor, akin to the wide range of visual styles in the cassette underground's mail-art scene. Grady remembers one pen pal who would write a given tape's song list on the back of old receipts. Others glued household objects such as aluminum foil onto tape cases. Grady herself was partial to sprinkling small bits of paper and glitter onto tapes, until she realized they could jam up a tape deck. Messages also emerged in the selection of music on a mixtape, in ways that often went beyond the already-powerful communication inherent in the music itself. Embedded inside sequences of songs could be hidden codes based on the styles of music, where the artists where from, what time periods they represented, and much more. Grady once received what she realized was a "message mix" after noticing that the first letter of each song spelled out the phrase "I like you." "I guess tapes were a way to express some things that you weren't ready to express ver-bally," she says. "The classic example of that is making one for someone you're trying to date or someone you're already dating."

Grady is quick to point out that the role of mixtapes that's often most emphasized in popular culture—romantic communication—is just one among many. Sharing musical knowledge, expanding on conversations, deepening friendships by connecting over songs that would then spawn connections to more songs: all these things make mixtapes a unique, indelible swirl of human interaction. "I think that's why I still love lis-tening to my own tapes that were made for me in high school and col-lege," Grady says. "You can hear your own progression through life, and through friendships, and friendships reflect who you are."

Mixtapes are literal conversation pieces, audio letters in which the songs could be taken as words, phrases, paragraphs, or even essays. They also can be timepieces, serving as snapshots, portraits, or even moving pictures. And just as any kind of communication can become a craft with its own conventions and innovations, composing a mixtape is often seen by practitioners as an art form. Every decision in the process could represent an aesthetic choice, a philosophical statement, even a treatise on the concept of communication itself. It all happens within a set of parameters and possibilities presented by the cheap, easy-to-use compact cassette, the canvas upon which mixtape painters apply musical brushes.

"If the mix taper values coherence—and coherence is especially favored among those who consider mix taping to be an art form—the selection of each song will be informed by at least five constraining considerations," posited Bas Jansen in the 2009 book *Sound Souvenirs*. "The song must fit the overall theme, continue the flow of the tape, be 'true of the mix taper,' take into account the tape giver's assessment of the musical taste of the recipient, and fit the predetermined time frame and two-act structure of the tape cassette. The result is a cohesive tape in which no song can be randomly replaced with another without destroying the structure."

Jansen's recipe for a coherent mixtape is just one of many, of course, as each mixtape maker continually chooses and redefines their own rules. But most people obsessed with making mixes work with some kind of plan, a set of often-unspoken guidelines that shapes the way they put songs on cassettes. "I was always trying to make the perfect mix," says Menghsin Horng. "To me, that meant each side was perfectly balanced, that both sides had tons of stuff that I loved and hopefully would be new to the recipient and that they would also love, and that the curation demonstrated exhibited my good feelings for the recipient."

To achieve those goals, Horng established her own way of making mixtapes that became a kind of ritual. She kept a "mix drafting notebook" with a page dedicated to each new tape. When the time came, she would open to a new page, pick a theme if necessary, and start filling paper with song titles and durations, which she would add together on her Texas Instruments graphic calculator. "I had a text file on my computer that listed the precise song lengths of every song in my collection," she says. "This data helped me plot out and calculate each side of a tape to minimize dead space." She'd swap and juggle songs on her list until she came up with the ideal sides, then record them onto tape from her turntable or her parents' five-disc CD player.

A mixtape made for Menghsin Horng by
her friend Kevin Potts, from the mid-1990s.
(Photo by Menghsin Horng)

In a final act of christening, she would listen to the whole tape by
herself while crafting the package's handmade art, before finally releasing it into the trusted hands of the postal system. "I would cut and paste
tiny little strips of white-on-black print for song titles, glued onto clip
art and glossy magazine images, the whole thing sprinkled with glitter
and encased in clear packaging tape," she recalls. "I wanted it to be clear
that each tape was unique, and a labor of love."

Mixtapes were similarly a labor of love for Gene Booth, who spent
nearly all his free time in high school and college making them. But love
wasn't his main motivation. "The idea of making a mixtape for a crush
has not often been my thing," he says. "For me it is about preaching the
gospel, really." Booth's tapes were about musical knowledge and connection, juxtaposing artists, scenes, and time periods that could inform one
another, as well as the tape's recipient. Sometimes this meant multiple
songs from individual artists, or even whole albums spiced with extras.
"I was big on making a statement for the bulk of it and then arguing

for it, and filling it out with context," he says. "So maybe put a Robin Hitchcock album on a tape with two Syd Barrett songs on the end. It was pedagogy, so I could take a person who hasn't heard this stuff to the next level. It's dialogic in that the two sides are talking to each other. It's context."

Even further context came from Booth's predilection for what he calls "what if" mixtapes. These were a kind of hybrid of mixtape and bootleg, focused on one artist whose outtakes and otherwise out-of-circulation recordings could form an important footnote to their history, or even an alternate history itself. For Booth, a "what if" tape was as much about seeking out material as it was about assembling it. "The story of mixtapes is a lot about dissemination of newly available information," he says. "Now no one understands that because we live in an era of box sets, in which everything has been rewritten and everything has been discovered." As examples, Booth cites tapes of the Beach Boys' at-the-time unreleased album *Smile* and Bob Dylan's *Basement Tapes*. He also notes fantasy-styled alternate-universe mixtapes that rearranged what corporations doled out, such as "an early 1970 tape, which took tracks from each Beatles solo record from that year to imagine a new last Beatles album." "My best of the 'Great Losts' was an Alex Chilton mix," he remembers. "I made the cover from a deeply distressed pic of Chilton on a rooftop I found in the *Village Voice*. Using copy machines at Kinko's and making a tape have similar energy to me."

When it came to music that was readily available, one of Booth's preferred mixtape formats included fifteen songs per side with one or two from some artists and five or more from others, to build an educational flow with links between works that he felt warranted equal attention. "I would always have one deep dive of one of them," he recalls. "I would think, 'You've got to hear all these Magnetic Fields songs or all these UK Kaleidoscope songs, but do you need everything by Melt-Banana or the Yardbirds?' I think choices like that make it so much richer in terms of making connections." There were no other limits besides what Booth could imagine—except for one. "It's gotta be a ninety-minute tape, and it's gotta be high bias," he says. "Fuck normal bias, and fuck anything other than TDK or Maxell, really. That's what it had to be."

For other mixtape makers, the art of mixing came from the immense possibilities that cassette tapes represented. This included the ability to marry any sound sources imaginable—not just music—into collages, not far from experimental tapes made in the 1980s cassette underground.

This idea first occurred to Jonathan Herweg when he was just eight years old, discovering his mothers' Magnavox dual-cassette boom box. "I would record myself talking as if I was the host of my own radio show," he remembers. "And then I would use the pause button to edit in real commercials and songs from whatever local station I happened to be listening to. I spent hours and hours alone in my room doing this make-believe radio show."

By middle school, Herweg was hooked on his local college radio station, WHRW at New York's SUNY-Binghamton. He scoured record stores for obscure sounds, taping them onto cassettes for mobile listening. Along the way he discovered all kinds of sources for mixtapes. One staple was a tape called *Henry Lee Lucas Confesses*, featuring spoken word from a notorious serial killer. Herweg mined it for sound bites, along with dialogue recorded from television shows, movies, poetry albums, and novelty records. "I would spend hours hunched over in front of my cassette deck trying to get the flow and the timing of each cassette just right," he recalls. "Slowly the sound bites started to take over and I would make more and more tapes with spoken word elements or a theme. At a certain point I started giving the tapes away to girls I liked or to kids I thought would appreciate them." Now a DJ on New Jersey station WFMU, Herweg doesn't make mixtapes much anymore, but he misses it. "I think it's a lost art," he says. "It was a labor-intensive love that was a personal obsession for years."

Like Herweg, Jeff Economy first discovered cassettes when he was young. Growing up in Chicago, he and neighborhood friends made recordings of themselves impersonating news anchors and reading made-up stories. An older sister turned him onto music, and soon he was checking out records from his local library and picking songs to put on mixtapes. "I knew nothing about wiring any components to each other directly, so I'd experiment with laying the condenser mic in front of the speakers until I found the right balance," Economy remembers. "I made pass after pass trying to get it just right, much to my mother's consternation."

At the same time, Economy recorded audio from TV shows and movies. "Having tapes that no one could get legally earned me some school-yard cachet," he admits. "The tape I made of the entirety of *Star Wars* by sneaking a recorder under a first-row seat got a lot of attention. 'Play that R2-D2 sound again!'" In middle school, Economy became obsessed with Dr. Demento's show of musical curiosities, taping episodes with a

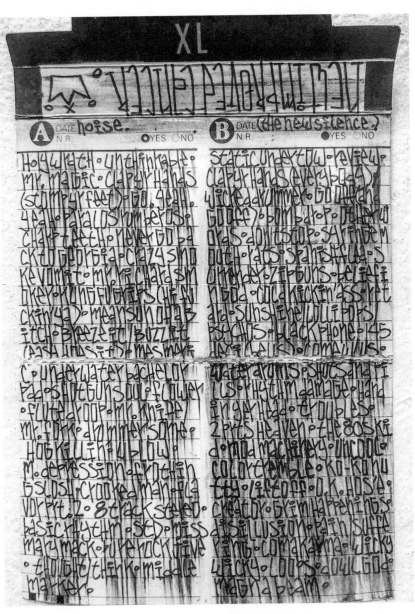

Cover art crafted by mixtape artist Jonathan Herweg. (Photo by Jonathan Herweg)

mic that he hung over his clock radio. Eventually he got better gear and jumped into making mixes obsessively, determined to have the weirdest and most eclectic creations among his growing group of tape-trading friends. "I spent countless hours working on my techniques, learning timings, getting the cueing just right, mastering fade-ins and -outs, making seamless transitions with nothing more than a steady hand on the pause button," he remembers. "I took it so seriously that I kept a hardcover notebook to record track listings and titles, where sometimes I would even record friends' reactions and ratings to take into account for the next one."

Economy's mixes were works of collage art, crossing music with found sound to show connections between artists and surprise with unexpected transitions. But he was also just as interested in the mixtape as courtship device. "To my mind the mixtape reached its apotheosis in the love mix," he says. "Done right, it was a perfect early gesture: love letter, soundtrack, Hallmark card, and handmade keepsake all in one. Putting together a mix that perfectly described (or engendered) a night of romance made you feel like you'd just out-composed Ravel's *Bolero*." He also extended his mixtape aesthetic to the packaging, at first using elaborate lettering with pens and markers, and later using dry-transfer Letraset (sheets of lettering that could be rubbed onto paper). He would mix typefaces and languages, add nonsense and in-jokes, and pore over old magazines looking for images to append. Like Herweg, Economy no longer makes mixtapes, having transferred his obsessions to his radio show on WGXC in upstate New York. And like Herweg, he misses it. "Making a whole contained little physical package gave me great pleasure," he says.

If a mixtape can be both art and conversation, can the medium be a message itself? After all, if two people want to communicate, all they have to do is talk to each other. Expressing something through a cassette tape introduces an intermediary, and conversing through music, sound, and collage adds layers of personal and cultural history and meaning. With that come opportunities for connection but also chances for mis-interpretation, on both ends.

"[The mixtape] conveys something of the character of the person making it, in theory, as well as being a display of commodity ownership, as opposed to copyright ownership, through creative juxtaposition," wrote Paul Hegarty in his 2007 essay "The Hallucinatory Life of Tape." "This investment (including when the tape is destined for your own use) means the mix takes on the character of a snapshot, and like Barthes' idea of the photographic image, it suggests narratives beyond it."

For some, these suggested narratives and added meanings can form a barrier between mixtape traders, one that is perhaps easy to hide behind. "It's a way of expressing oneself without pretending that there is some honest, unmediated 'self' to express," asserts Kamal Fox in a 2002 essay on mixtapes. "It is yet another way of saying 'I think you are really neat' without being cliché or falling into the whole trap of sincerity. It is a cultural product made up of cultural products, a referencing system that references other such systems. Since all musical practices are based on this borrowing and quoting and this wish to build social bonds through cultural commodities, mixtapes, in turn, are not about the disclosure of the 'self,' they are instead about creating a sense of availability."

In his entry in Moore's *Mix Tape: The Art of Cassette Culture*, critic Matias Viegener agrees. "The mix tape is a list of quotations, a poetic form in fact: the cento is a poem made up of lines pulled from other poems," he writes. "Similarly an operation of taste, it is also cousin to the curious passion of the obsessive collector. Unable to express himself in 'pure' art, the collector finds himself in obsessive acquisition." Adds Dean Wareham in that same book, "There is something narcissistic about making someone a tape, and the act of giving the tape puts the recipient in our debt somehow. Like all gifts, the mix tape comes with strings attached."

These are all fair points, though they are also a bit cynical. Perhaps tributes to the communicative power of mixtapes are slightly naïve in suggesting that direct connection can come through music made by a third party and imprinted onto a physical object. It's a message the recipient can make all kinds of assumptions about and inferences from—as well as being free to listen to as little of it as they'd like, whenever and wherever they'd like—without much clear input from the maker. But for some, everything that the mixtape and its contents puts between two people is part of the magic, much the way music itself is not a one-dimensional statement but a whole package of meanings tied up in histories and allusions and changing social conventions. When you make someone a mix, maybe you want to tell them something, or maybe you want them to discover something even you didn't know when you made it.

Even the sheer physical nature of the mixtape is something that, instead of getting in the way, can serve as connection. This helps explain why the cassette tape is a more celebrated and adored way to share musical mixes than CD-Rs, hard drives, or streaming playlists. Like humans and their interactions, the cassette tape is malleable, fragile, unpredictable, and idiosyncratic. It can change, it can decay, it can be repaired. It can even feel like an extension of the bodies between which it passes. "Listening to cassette tape music, it seems more relative, in a way, to the human condition, because our bodies aren't digital; we're not robots," says Thurston Moore in the film *Cassette: A Documentary Mixtape.* "And your body feels different every day, it goes through changes, it gets worn down the more your body goes over that tape head of life."

Beyond their physical nature, mixtapes have undoubtedly been conduits and catalysts for community building. Ask almost any devout mixtape trader and you'll find that they've forged long-term friendships based on trading tapes, even with people they've only met through the mail. "I definitely learned a lot trading such mixtapes with informed collectors," says Steven Krakow, a Chicago journalist and music historian. "I learned even more through a network of tape traders of old psych/prog/avant rock stuff, who'd make lists of what they owned for trading ease." Krakow still runs a mixtape club, in which he sends out six tapes a year of "rare sounds, with themes like glam, psych, punk, hard rock, acid folk."

Fred Church literally made a community with mixtapes, forming a club with two friends who have collaborated on mixes for decades. "The customizable format—where you could create your own content and grab songs from here and there—was a powerful idea," Church says. "Making mixtapes as a group activity was logical, because we hung out and played each other songs all the time, as part of our daily lives. When we were making a featured group concept tape such as *Chaotic Nuisance from the Flaming Snatch of Humanity*, it would be a bit more considered, as we were specifically trying to make something cool. But we trusted each other's taste, and each of us would propose songs, and we'd feed off the vibe."

Stories like these have emerged continually since the advent of the cassette tape. In 2003, the Museum of Communication in Hamburg, Germany, decided such mixtape narratives were worthy of public exhibition. Its project *Cassette Stories* featured tales from eighty different mixtape makers on what made the practice so meaningful and so obsessive for them. Traces of the exhibit seem to have faded (only one article about it came up in an internet search), much the way millions of mixtapes have been tucked away in attics, buried in basements, or even thrown into the trash. But the concept behind them, and what they have meant for human musical connection, remains indelible.

Sarah Grady doesn't make mixtapes much anymore, but she thinks about them all the time. For her, they are living documents that still have something to say to their makers and receivers, even years removed from their original contexts. "Just today one of my friends sent me a bunch of pictures of mixtapes he found that an old girlfriend of his had made him," she says. "And he was talking about how he's listening to it now and thinking about his role in the dynamic of the relationship that he can actually hear in these mixtapes that he didn't really pick up on at the time." In 2016, Grady's ongoing obsession with mixtapes led her to create the blog *Made You a Tape*, where she shares, as the site puts it, "stories and interviews about the mix tapes that meant so much to their recipients." Often this involves reuniting two people who shared mixtapes, hearing their reflections on how those interactions began and what effect they have to this day.

In one entry, she reconnected her friend Michael Honch with legendary punk musician Ian MacKaye of the bands Minor Threat and Fugazi and the iconic label Dischord. In 1987, Honch wrote to MacKaye asking if he had any recordings of Washington, DC, band Dove, and MacKaye sent back a tape that included them and four other bands, including some of Fugazi's first recordings. A fastidious collector and cataloger, MacKaye still has Honch's original letter, and the meetup spurred him to remember how important mixtapes were to him in the early days of punk.

"I've been typing up my 1984 journals, and I write that Henry [Rollins, of the band Black Flag] was visiting, but spent all of his time sitting upstairs making tapes," MacKaye said. "'Cause that's all we did. You'd go to someone's house, you'd make tapes because you were getting back on the road. . . . Everyone had stacks of tape decks and were making tapes. That was the discourse." "One of the things that impressed me about this so much was the generosity of [you sending me the tape], without expecting anything in return, except to listen," Honch said to MacKaye. "It's something that I carry with me. It's one of those ripple effects. I think about ways in which this changed how I think about sharing work with other people, that it's possible to communicate with people in ways in which there wasn't an economy to it."

In another entry on *Made You a Tape*, Grady reached out to her old mixtape pal Elke, who painstakingly documents her creations by writing lists of songs on index cards that she stores in a recipe box. She still refers to these notes when making new mixtapes, so as not to repeat songs for specific recipients. Her notes work in the other direction too: before making a mix, she usually sketches out a worksheet of what she plans to put on the tape, adding up times to make sure she maximizes each side, then updating the worksheet once she puts the mix together. For Elke, the act of making a mixtape is intimately tied to friendship. "It's cool to think that the person could be experiencing these songs that you know intimately . . . for the first time on your mix," she said. That also makes mixtape trading a risky, vulnerable act: If the person doesn't like what's on the tape, are they in some sense rejecting you? "It's like a snapshot of yourself and what you're listening to at that time, and you wanna get that down for this other person too that your life is being shared with," Elke continued. "The mix is your product even if it's made out of other people's music."

In another entry on the site, Grady's friend Mat Darby concurred: "I think there's always sort of a self-consciousness of making a tape and saying, 'Okay, this is not just the tape. This is part of my being, and I'm giving it to you.'" For Grady, the positive side of this sharing outweighs the dangers. "To listen to a mix tape is to hear one person's personal take on the high notes of music history—the actors and events that impacted her," she wrote. "She is the author of her own personal history lesson, and you are the lucky pupil. I'm not sure that there is a better way of learning about the world than this."

There also might not be a better way of learning about yourself through music than making mixtapes. Picking what music matters enough to you to share with someone can be an act of self-definition, or even self-realization. "Making mix tapes is really . . . a way to experience yourself," Jeff McGrath said in conversation with his friend Eric Hatch on *Made You a Tape*. "It has something to do with understanding our relationship to our thing we love. It's a way to spend time with it or meditate with it . . . and that's sort of nourishing. Because it's just you and the music."

Erin Margaret Day knows a lot about the solitary aspect of making mixtapes. "I spend so much time with each tape, just listening to it over and over again," she says. After buying a new tape deck in 2020, Day began a series of seasonal mixtapes. Four times a year, she creates a mix based on a theme, sends the tape to one friend, then posts a digital version of the audio on the internet for anyone to hear. "Some songs can just be there for references to the season," she explains. "Others can establish that sense of what it looks like and feels like outside. Like, I don't know if *Power, Corruption & Lies* by New Order has any direct lyrical content relating to winter, but it's very much a winter record for me." She gives each mix a name—her winter 2021 entry was titled *Recorded Syntax*, taken from a song by the band June of 44—crafts her own artwork for the case, and provides a handwritten list of the bands and tracks she includes. "*Recorded Syntax* also seems like a great metaphor for what crafting a proper mixtape is," she wrote on her website. "I record tracks syntactically, to conjure feeling, narrative . . . often even crafting arguments, playing with tensions both sonic and lyrical."

Born in 1987, Day is not new to cassette tapes. As a kid, she recorded songs off the radio to learn how to sing them, sometimes pilfering blank cassettes from her stepfather's stash. In her teenage years, inspired by communities she found on the social-blogging site LiveJournal, she started making and trading mixtapes. "I loved writing things in different ways and experimenting with making my letters," she recalls. "What I was really interested in was coming up with little challenges for myself. I liked using opening and closing tracks from albums in the opposite way, and I got into wild transitions, putting effort into joining sounds that you wouldn't ordinarily expect to hear."

Over the course of her mixtape journeys, Day learned a lot about music and made some good friends. One of them, whom she had corresponded with across states, became her roommate when they both ended up moving to Chicago. In her twenties, without enough income to invest in tapes and records, she turned her mix inclinations to the streaming music application Spotify, making playlists with the same attention to detail as when she made mixtapes. The seasonal mix idea started there, but she eventually became disillusioned with streaming music and the costs involved. For one Spotify playlist, she wrote song titles out on paper, as she puts it, "just to remember how much I love doing that. And I'm glad I did; it led me not just back to making mixtapes but to doing things that bring me joy again generally."

Now that she's making cassettes again, the rewards Day first found in the practice have intensified. "The seasonal mixtapes have taken on this sort of magical quality," she says. "When I was working on my spring mixtape, every day that I worked on it, the sun would come out and all the snow would melt. It's like I'm starting to control the weather with my stereo, through spending so much time on each of these and reflecting on my own life and putting what I want to manifest into each thing. It's become part of a process of self-actualization."

In 2017, British artist Mandy Barker took a vacation in Fuerteventura, a Spanish island near the northwest coast of Africa. As she walked along the beach, she came across a cassette that had washed ashore. Barker's art deals in part with environmental concerns, specifically with the way we discard and waste plastic products, so the tape was a good fit for her work. She found an audio professional who could restore the music from the tape, and incorporated that and a list of the songs into a piece she called *Sea of Artifacts*.

Two years later, a Spanish tourist named Stella Wedell was traveling in Stockholm, Sweden, when she happened upon Barker's exhibit at a gallery. The list of songs on the tape—a diverse lineup that included reggae, dance music, and Disney tunes—seemed familiar to Wedell. She snapped a photo and took it home, where she still had some CDs based on mixtapes she made in the early 1990s. One of those CDs had the exact same track list as the one Barker had found, and Wedell realized that the tape was actually hers. She had lost it when visiting the Spanish island of Majorca, more than 1,200 miles from where Barker discovered it.

"I always made tapes from my CDs at this time to listen to them with my Walkman, especially for holidays," Wedell told CNN. "To think that a tape I could have lost more than twenty years ago had been found was incredible." "Even after getting the tape to play, it was then an astounding chance for Stella to walk into my exhibition and recognize her tape," Barker told *Sky News*. "She said she was shocked to find it, and when I read her email I couldn't believe it either. It was shock all around."

Whether they travel across continents and years or simply from one room to another, mixtapes have always had a unique capacity for communication—both from the music to the maker, and the maker to the listener. In this way, the cassette tape birthed a new kind of conversation, forming connections that were mediated in a potentially powerful way, because the medium could become integral to the message. "I have not encountered a technology for recorded music whose physics are better suited for fostering the kind of deep and personal relationships people can have to music, and with each other through music," wrote Nick Sylvester in *Pitchfork*. "It whittles down our interactions with music to something bare and essential: Two people, sometimes more, trying to feel slightly less alone."

CHAPTER SEVEN

Tape's Not Dead

2

SXC 6365

MADE IN ENGLAND

100   50   0

The Cassette Comeback

"Most people would probably think there aren't 100,000 cassettes left in the world," said Steve Stepp. "[But] I've got an order of 87,000 going out today." Talking to *Rolling Stone* in 2016, Stepp explained how his Missouri business, National Audio Company (NAC), thrived by doing what it had always done: making tapes. He and his father founded NAC in 1969, just a few years after Lou Ottens and Philips created the compact cassette. Initially, NAC made cartridges used primarily by radio stations. By 1980, they had shifted to manufacturing cassettes, not long before tapes began to outsell vinyl in America.

The cassette tape's dominance didn't last long. In 1983, Philips launched a new digital audio format developed by Ottens and his team, promising more pristine sound than found on analog media. They called it the compact disc, hoping that first word would ring in consumers' heads due to the success of the compact cassette. Within a decade, the

CD outpaced both tape and vinyl in American sales, and many duplication firms shifted focus accordingly. By 2000, even Philips itself stopped manufacturing cassettes. But all along, NAC continued to make them, even buying equipment discarded by their competitors. As orders for prerecorded music cassettes dwindled, NAC relied on audiobook orders as well as institutions—government agencies, schools, libraries—needing duplication of spoken word and educational tapes. One Christian group in Missouri ordered 250,000 blanks weekly, using them to record and distribute sermons. "We didn't know the cassette was dead," Stepp told *Rolling Stone*. "We were told it was dead and we never believed it."

In the late 2000s, Stepp's convictions paid off. A few years before, vinyl had made a comeback, initially via independent labels whose fans preferred physical media to digital downloads, and later through major labels seeking an upscale market for nostalgia-tapping reissues. But as vinyl resurged, it also got more expensive and time-consuming to make. The few active pressing plants became overwhelmed, thanks in part to an annual event called Record Store Day, which flooded the market with limited-edition LPs sold exclusively in brick-and-mortar shops. In response, some indies turned to cheaper, faster cassette manufacturing, which cost about two dollars per copy compared to at least five times that for LPs. From 2009 on, NAC's sales jumped 20 percent per year; in 2014 they made 10 million cassettes, more than in their 1980s heyday. Seventy percent of those were prerecorded music tapes.

Historically, this cassette comeback was relatively small. Sales remained far below 1980s and '90s peaks, and vinyl was more widely coveted. Additionally, the increasing popularity of downloads and, later, streaming audio took a huge toll on physical media sales. But that also became a motivator for those turning to cassettes. They saw tapes as an alternative to the algorithm-influenced listening of services such as Spotify and YouTube, their low financial returns for artists, and the sterile, impersonal experience of listening to music on a computer or a phone. "There's a dirty little secret here," Stepp said to the *Springfield News-Leader*. "We've got an advantage. Your ears are analog. The world is analog."

NAC continues to flourish by sticking with reliable methods and adapting to new challenges. Much of its duplication machinery was made in the 1990s or earlier. Their apparatus that shrink-wraps tapes dates to 1938; it was originally used to encase cigarette boxes (oddly apt, since audiotape technology itself came in part from Fritz Pfleumer's repurposing of cigarette paper). In the late 2010s, when most companies that make the materials for tape had closed, NAC constructed a new, fifty-five-ton machine to create their own tape stock to the tune of 20,000 feet per minute.

In 2018, Stepp estimated that he was working with about 3,400 labels worldwide. That list wasn't made of just independent companies. It also included majors who, just as they did when vinyl resurged, jumped on the cassette trend. When Pearl Jam put together a tenth-anniversary box set of their Epic Records album *Ten*, they included a cassette of the original demo, ordering 15,000 copies from NAC. Stepp's company also made 25,000 cassettes of *No Life 'til Leather*, the legendary 1982 Metallica demo tape that kick-started that group's success. That reissue resembled a homemade 1980s dub, with a faux-handwritten cover mimicking the look of a fan-made tape.

Nothing symbolized the major-label rush toward tapes more than *Awesome Mix Vol. 1*, a 2014 cassette version of Disney's chart-topping soundtrack to the film *Guardians of the Galaxy*. Featuring songs that one of the movie's characters listens to on a mixtape, it was the first cassette released by the entertainment conglomerate in over a decade. "We had an order in [at NAC] and it was going slowly," remembers Doug Kaplan, co-owner of Chicago tape label Hausu Mountain. "I called my rep there, and he put me on the phone with the dude who runs the place, who was like, 'I'm really sorry, Disney bought out our entire production line for days to do the *Guardians of the Galaxy* cassette.'" For label runners such as Kaplan who had made, sold, and listened to tapes for a while, the cassette had become, in one sense, too alive and well.

The death of the cassette tape has been predicted, feared, and pontificated about for decades. As early as 1980, before CDs were even on the market, a *New York Times* article predicted that coming digital technology would render analog formats obsolete. Even as the cassette remained popular well into the 1990s—the *International Herald Tribune* noted that in 1998 it was still "the world's most widely used sound recording and playing medium"—the assumption that it would die became so ingrained that signs of life were often taken as a surprise. The feelings in the air were best expressed by tape aficionado Scott Marshall in his 1995 requiem "Long Live the Humble Audio Cassette: A Eulogy." "The Future looms upon us now—and brothers and sisters, like it or not, it's in binary code," he wrote. "Most aspects of 'modern' life will exist for the foreseeable future, at least on some level, as digital data storage. . . . But let us hope there will always be time for analog circuitry and audio cassettes."

The cassette certainly did decline in popularity over the next two decades, and proof abounded. "Not Long Left for Cassette Tapes," declared the BBC in 2005, noting that prerecorded tape sales in the UK had dropped from a 1989 peak of 83 million to just 900,000. The *Metro* wrote under the title "Death of the Cassette Tape" that UK sales of blank tapes had gone from 95 million in 1990 to below 1 million in 2007. In 2008, the *New York Times*, under the headline "Cassette Tape Going the Way of the Eight-Track," claimed that sales of tapes had dropped to 400,000 from a 1997 figure of 173 million, and sales of portable tape players had dwindled to 480,000 from a 1994 high of 18 million. (The *Times* author couldn't help taking a shot at tapes as well: "While vinyl records have always been prized artifacts for their devotees, the plastic cassette tape has little sex appeal.")

More elegiac notices proliferated too. In 2007, the *Independent* ran an article called "Farewell to Cassettes: Tales of the Tape," in which fans mourned the cassette in romantic terms. "It's hard to feel nostalgic about cassettes because they were so awful," admitted *Mojo* editor Phil Alexander. "But when you see pictures of an old TDK tape with handwriting on it you can't help feeling warm. It just shows you, nostalgia can creep up on you at the most unexpected times." Alexander was on to something. Despite the cassette's decline in popularity, it stuck in people's minds, and its death never came to pass. Some once-crucial aspects did perish, for sure. Take the car tape player, once a major factor in the cassette's success. By 2010, the only company still offering them was Lexus, who stopped just a year later.

Still, a cassette comeback started even before elegies such as the *Independent*'s emerged, and in a few years, the press caught up. "Can Cassette Tapes Be Cool Again?" asked CNN in 2013. "Cassette Tapes Are Almost Cool Again" answered *Vice* that same year. In 2014, *Newsweek* queried, "Is the Cassette Renaissance for Real?" while in 2016, *Rolling Stone* explored "Why the Cassette Tape Is Still Not Dead." Revivals in non-Western countries such as Japan, China, and Indonesia garnered attention as well. In 2016, *Vice* profiled Amar, whose one-man tape duplication shop in Indonesia had thrived in the 1980s by making pirated tapes. Despite drops in business (and a few police raids), he hung on, and by the early 2010s labels came to him to order authorized copies of their new releases. "I know all these people by word of mouth," explained Amar, who now makes 2,000 tapes per month. "I don't do pirated stuff anymore. I'd rather do this and help out these musicians."

Unsurprisingly, a backlash to the cassette comeback came along too. In 2009, just two years after the *Independent*'s eulogy for tapes, *Pop Matters'* Calum Marsh wrote, "At best, the cassette revival is merely a vacuous fad of no genuine value; but at worst, it's a confused, regressive cultural misstep more dangerous than most would care to admit." In a 2015 *New York Times* piece, Rosencrans Baldwin was more blunt: "As a format for recorded sound, the cassette tape is a terrible piece of technology. It's a roll of tape in a box. It's essentially an office supply." "I'm actually quite amused by the audacity of anyone attempting to drum up some sense of nostalgia for a format that was barely tolerated in its supposed heyday," *Pop Justice* editor Peter Robinson told the *Guardian* in 2019. "It's like someone looked at the vinyl revival and said: what this needs is lower sound quality and even less convenience. I think labels know full well that almost every cassette they sell is going straight on a shelf as some sort of dreadful plastic ornament."

To skeptics such as Robinson, the cassette revival was about escapist nostalgia or blatant trendiness. In other words, an attempt to look retro-cool without caring about the music, which, conventional wisdom held, sounded better on every other format. On a mainstream level, these charges had some merit. Disney's cassette version of the *Guardians of the Galaxy* soundtrack was probably a stab at cashing in on nostalgia, since a corporation that large surely didn't need to save money by making tapes instead of vinyl records. But for small labels, financial concerns mattered, and if nostalgia helped make their releases viable, that was perhaps a welcome byproduct but unlikely to be a motivation.

Besides, nostalgia for cassette tapes isn't necessarily an escape, a way to avoid the reality of the present by running to comforts of the past. Nostalgia can be about finding value, or even advantages, in old things applied to the here and now. The cassette's best qualities—cheapness, accessibility, compactness, user control—are traits that no single format has combined in exactly that way since. Tapes symbolize a time when autonomy and affordability were perhaps more integral to music fandom, even if that time was during some current fans' childhoods. "Cassettes are a perfect medium for children," wrote Sarah Renberg in the magazine *Basic Paper Airplane*. "They are handy, inexpensive, easy to operate, and well-sized for small hands. My sister was able to buy a tape with her allowance, and I was able to flip the tape over and press play. When I say this was a form of magic, I mean it was very accessible magic. Its availability to us was part of its power."

Debates over pragmatism versus nostalgia point to bigger questions about the resurgence and resilience of the cassette tape. Can its use be a revolutionary statement, perhaps even a consequential political act? Early on, cassettes freed musicians and listeners from the expense and limits of industry-controlled formats. Now, cassettes can offer a way to avoid corporate streaming services, whose offer of listener freedom is a bit of a mirage, considering the algorithms that push them toward specific artists, gather their personal data, and subject them to advertisements. Cassettes can also provide a more intimate way to share music with others. Giving someone a handmade mixtape is surely more personal than sharing a playlist, whose creation is more akin to data entry and which is usually accessible only through paying for subscriptions or enduring ads. Cassette labels are also, hopefully, less likely to shortchange artists for their work, while streaming services offer fractions of a penny per play.

The politics of cassette tapes can even be audible in the music. After all, the fact that the musicians and labels who use them are often supporting sounds and styles that have limited commercial appeal is a big part of why they turned to cassettes in the first place. The hissy, distorted, unpolished tape is a thumbed nose at professionalism and a statement about what art can and should be available. "The physicality of cassettes and the music on them represent a new kind of protest music in a post-9/11 world," argued Craig Eley in his 2011 essay "Technostalgia and the Resurgence of Cassette Culture." "By sonically bathing their music in an aesthetic of uncertainty, distance, longing, and mechanization that characterized moments of the Cold War, tape artists use technology and technostalgia to express their ambivalence toward dominant music culture practices as well as global politics. In an increasingly digitized culture, the physical becomes political."

The resurgence of the cassette tape also brings up questions about how we view the past, how the past viewed the future, and how these things mix in the present. Finding value in technology some consider obsolete questions the whole idea of obsolescence and whether the things we discard are as fixed in their form and function as we assume when we categorize them and assign them to time periods. In a 2017 issue of the journal *Twentieth-Century Music,* Joanna Demers called this process "creative anachronism," or "an example of a creation rather than a melancholic reiteration." "Artists are smart enough to know that the past is permanently off-limits, if we treat it as a guiding principle that must be copied to the smallest detail," she wrote. "But if elements of the past, such as 1980s tape culture, are used as starting points for fictions, the tenor of the activity changes to burgeoning creation. Anything is possible in such moments of creative anachronism, especially when that which is created is patently impossible." Demers's point is a bit intangible, but it highlights how the cassette comeback isn't a simple binary of past and present. Instead, it's an intermingling of the two, perhaps even a redefining of what tapes mean to us now and what they meant to us then.

Another important rethinking comes in the response to a criticism raised above by Peter Robinson. He suggests that, because cassette players are now relatively scarce, tapes don't just devalue music through their inferior sound quality. They actually make music get ignored completely, as the tape becomes a hipster decoration, a wall hanging from the past rather than a functional object in the present. It sounds like a fair point: if the music isn't listened to, a cassette tape is in fact nothing more than a totem. But for dedicated, practicing cassette labels and their audiences, that's not how it works. Quite often, cassettes come with a download code, via which the buyer can access a digital version of the music to store on hard drives and portable digital players. This gives fans the opportunity to support an artist at a much higher rate of remuneration than listening to them on streaming services, as well as the chance to own something more concrete than a file made of ones and zeroes.

Researching cassette culture in the early 2010s, Iain Taylor found that some of his interviewees only listened to digital downloads yet still valued the cassette tape beyond its retro qualities. "Participants emphasized the cassette as a convenient and cheap way of catering for their desire for a physical thing," he wrote. "The majority of them highlighted that even though they were aware that they could access the same music online for free, the purchasing and ownership of a certain cassette is important as a means of putting money in the hands of the artists and labels that you enjoy. It becomes a symbolic token of your involvement with a scene, as a person who doesn't merely observe or listen, but instead is an active participant in independent music who actively contributes to the production of culture." As one of his subjects put it, "No one wants to buy a digital download code at a show. You want to come away with a thing."

Taylor's research led him to conclude that cassettes have become a "hybrid artifact." Owning a tape was, as he put it, "not about fetishizing the cassette as a retro artifact, but about redefining it." Even if the cassette has been somewhat stripped of its utility in music listening, that doesn't mean the object itself is strictly symbolic or nonfunctional. Instead, it has become a kind of mentally charged conduit between listening to music and supporting artists, between the ephemeral experience of digital signals turned to air and the tangible relationship that music fans have with bands, labels, and even the physical spaces in which they all interact.

By breaking down assumed dichotomies between form and function, style and utility, Taylor points to an essential reason why the cassette tape has survived and thrived. It's been adopted and adapted, not resurrected as much as repurposed, recast into new roles that draw meaning from their reimagining of the past and their disruption of the present. "By viewing artifacts like the post-millennial cassette not as a materialistic fetish, but rather, as a symbolic thing, linking together sets of digital and physical cultural practices," Taylor concluded, "we can better understand the richness and totality of how these practices as a whole enact meaning and value, without over-privileging or arbitrarily dismissing one or the other."

Taylor's point resonates with at least one current cassette label. "That's where we're at generationally," says Doug Kaplan of Hausu Mountain. "We have memories of life pre-computer and post-computer, and we're in this place between physical and digital worlds." "It's a weird obstacle to overcome because of this perceived throwback or retro association with the cassette," says Kaplan's label partner, Maxwell Allison. "We don't care about that at all. We're not doing this to evoke a golden past. It's more that everything is still valuable."

Running Hausu Mountain since 2012, Kaplan and Allison are part of an informal, interconnected global community of small labels focused on cassette releases. Spread around the globe, these DIY companies— usually run by one or two people—often have creative names to match their aesthetics, such as Moon Glyph, Not Not Fun, Unifactor, Dinzu Artefacts, and Tsss, among scores of others. Since around the mid-2000s, these imprints have bypassed the dominance of digital downloads and streaming without the kind of financial wherewithal needed to press vinyl. Unlike home tapers in the 1980s cassette underground, these labels don't usually hand-dub their releases. Instead, they get them manufactured at places such as NAC and sometimes distribute them through the same channels that carry vinyl and CDs.

The music supported by these tape labels often pushes the boundaries of convention, blending lines between experimentation, noise, improvisation, outsider art, and electronica. Commercial appeal is rarely the point, freeing artists to explore new ideas without the pressure of recouping investments. "The big labels have many resources," explains Daniel Castrejón of Mexican label Umor Rex. "But also many limitations in the creativity, because they want to please." "People willing to buy tapes and support tape culture are already more open to new experiences than the general populace, most of whom would not pick up a free tape up off the street," adds Dylan McConnell of the Field Hymns label. "You can get more creative and push the boundaries of what is acceptable with an audience like that."

Such mysterious, unconventional sounds match perfectly with the analog reproduction of cassettes. Hiss, distortion, and playback anomalies are welcomed as contributors to the unpredictable nature of the music. "We definitely think about how things will sound on tape," says Dwight Pavlovic of West Virginia's Crash Symbols. "And we love music made with that in mind, regardless of how it sounds." "The limitations of the format can have a positive impact on the music, especially in music, where there are few rules," adds Paul Condon of the Irish imprint Fort Evil Fruit. "It can make for a good framing device and encourage more thought and consideration than a digital-only release might be subject to."

"The cassette offers the exact artistic and commercial freedom that most jazz musicians look for," says Nate Cross of the jazz-leaning label Astral Spirits. "It mirrors a time back in the sixties and seventies when vinyl pressing was cheap and there were lots of people self-releasing albums. There wasn't any pressure to create this artistic masterpiece— it could just be a capture of a certain point in time. Cassettes give that freedom back to a lot of jazz musicians." "You can try new things, release work anonymously, whatever is necessary to let the work speak on its own terms," says Scott Scholz of Tymbal Tapes. "It centers the relationship around art and not commerce, even when there is a bit of commerce involved—but let's be honest, the only people actually making money on cassettes are PayPal, [online retailer] Bandcamp, and postal services around the world."

Many of these cassette labels value not just music but artwork, packaging, and overall presentation. Taking advantage of the size, shape, and fold-out potential of cassette covers (known as J-cards), these imprints forge visual identities to accompany their sonic signatures. "Music is a wider experience than just what you hear," says Eamon Hamill of Spain's Strategic Tape Reserve. "Ideally, the album art, concept, liner notes, and general atmosphere should really be integral to the music itself—part of the listener's experience of the music." Some labels establish aesthetics that run through all their cassette covers, often via consistent use of fonts and layouts that form a kind of template, be it obvious or subtle. Others apply wildly different styles to each release, with the common denominator being the intent to catch eyes and minds. "The goal of the cover art," explains Steven Ramsay of Constellation Tatsu, "is to spark the imagination right before the music hits."

For Adam Svenson of the Seattle, Washington, label Eiderdown, visual art has always been a motivator. "When I started up the label, it was partially with the intention of making sure that my friends that did interesting weirdo artwork might have a new platform to showcase their art," he says. "Honestly, I was tired of how lazy packaging and art was getting with record labels. If there's no story there and it looks like no one cares, why should I bother?" So Svenson called on Seattle artists Max Clotfelter and Aubrey Nehring, both of whom make hand-drawn art that feels oddly alien. Strewn with bright colors and intersecting shapes, their covers often resemble monster cartoons melted into abstraction, giving Eiderdown what Svenson calls "a sixties-underground-comics-meet-bootleg-psych-record vibe."

Sometimes, the designers of cassette covers are the people who run the labels themselves. Field Hymns' McConnell makes art that varies from homages to blank tape designs to melding of sci-fi and occult imagery. "I am way too ADD to conform to structure," he admits. "I need the flexibility to describe the music accurately, or at least through my own lens of constantly changing influence." At Orange Milk, co-owner Keith Rankin uses pen and ink, cut-and-paste collage, and Photoshop manipulation to create art that feels both futuristic and retrospective. Pointing to the cover of a release by his own musical project Giant Claw, which he made in collaboration with painter Ellen Thomas, Rankin says, "I love how it combines surrealism, classicism, and conceptual undercurrents,

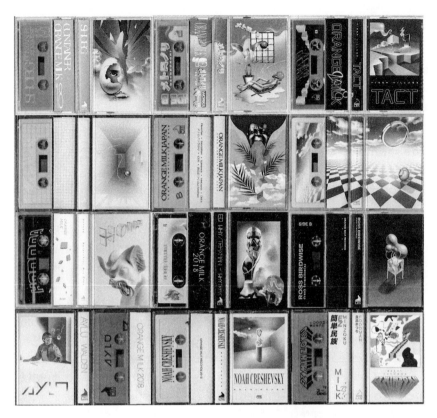

A set of cassette releases on the Orange
Milk label, designed by co-owners Keith
Rankin and Seth Graham as well as artists
Biata Roytburd and Ellen Thomas.
(Courtesy of Orange Milk)

and how each of those elements enhances the other when probed." For
Umor Rex, label head Daniel Castrejón builds visual themes through
a geometry of words, in what he calls "a game between lines, forms,
and space." "I try to avoid too much information; it's always a process
of reduction," he explains. "The cassette is a rectangular space, so you
can do a vertical design or a landscape. It is a natural form for artwork."

Umor Rex often releases its titles in sets of two or three, what labels
often call "batches." Besides helping streamline the duplication process,
the practice of releasing in batches leverages a label's track record of
finding good music, hoping that will entice fans to acquire them all. "It
encourages the adventure of simply trying new things, sometimes a lost

art in music listening," says Scholz. "And it can be a catalyst for community around the music, where listeners begin to trust the curatorial intuition of their favorite labels."

Many of these labels rely heavily for sales on the online store Bandcamp. There, they can offer batches at a bulk-rate discount, and listeners can download digital versions immediately after purchasing physical tapes that arrive later in the mail. At a time when stores and distributors are shrinking or disappearing, Bandcamp has become a boon for tape labels, enabling them to sell directly to customers. It also has an editorial section where writers recommend releases and profile artists and labels (I am a contributor; for two years I wrote a monthly column about new cassette titles). Bandcamp is a tech company that began independently, then was purchased by a video game maker, so its priorities could certainly shift in the future. But for now, the vast majority of tape labels appreciate the website that has helped them stay in business.

As more independent tape labels have emerged, they have overlapped in ways that have formed a kind of virtual cassette neighborhood. Some musicians have releases on many of these labels, while the labels themselves are often run by musicians who release music on other labels as well. Collaborations are common, as are trading tapes and creating new projects together. "Immediately after starting the label, I began to feel part of an international community," says Joshua Taibba of Los Angeles–based label Already Dead. "I don't know of many other fields or industries like the cassette world that are largely void of competition and focused more on creating a culture together." "We are always keeping up with labels directly, putting artists with demos in touch with other labels if they aren't a good fit for us, and supporting each other," says Bobby Power of Georgia's Geographic North. "Some of these online relationships have bridged into actual meetups and hangouts." "At the end, a common factor between the small labels is that all the people involved are music collectors," says Castrejón. "I'm not sure if that happens in corporations."

Because they're primarily working outside of mainstream channels, many involved in cassette tapes today view outsider interest in their dealings with skepticism, especially when it comes from the corporate music industry and press. Ask what they think about articles that proclaim "The cassette is back!" and you're unlikely to get the same enthusiasm in return.

Mike Haley, *Tabs Out* cassette podcast: "It's very silly, but expected. Obviously, cassettes are not coming back as a viable [mainstream] format. That is absurd. But ya gotta make content, I guess."

Maxwell Allison, Hausu Mountain: "It's funny how it spreads so slowly and stupidly from pundit to pundit. Someone [in the media] will discover that they're back, because they didn't just Google it and see that there's been ten years of these articles."

Ryan Masteller, writer: "It's laziness, journalistic tourism, like, 'Oh, here's this neat thing I forgot about, some people are still grinding away at it!' It's not a question of those articles elevating the cassette to a level of something that 'matters'; it's the joke of journalism's fickle attention span latching onto a shiny bauble and pronouncing it an objet d'art."

Dave Doyen, *Tabs Out* cassette podcast: "What I have read all has basically the same boring capitalist frame. In reality, there has been a pretty active cassette community for decades, and I'd be much more inclined to read articles about the people who have kept that community alive than how cool it is to have a Walkman now."

Paul Condon, Fort Evil Fruit: "With major labels getting in on the act—which has a lot more to do with gimmickry and nostalgia than the underground use of the format—it's reaching saturation point. The perennial 'tapes are back' article now conflates and confuses DIY activity with Jay-Z albums being put out on cassette and quotes industry sales figures that don't have anything to do with the former."

Scott Scholz, Tymbal Tapes: "Folks get queasy at the notion of those 'cassettes are coming back' think pieces, for sure, but there is some truth to it. . . . Perhaps it's more important to hope that what cassette and vinyl renaissance moments really mean is that people are finding ways to connect through music that's less mediated through corporate gatekeepers."

Dwight Pavlovic, Crash Symbols: "I've definitely had moments where I'm on the 'don't call it a resurgence, we've been here for ages' team, but I also think there's something useful about seeing that idea recontextualized or updated frequently."

Michael Potter, Null Zone: "There are tons of folks who don't even consider the cassette a viable medium. So even if some are only buying cassettes to keep as a type of artifact and never listen to them, I am glad for any support we get. . . . Everyone I know doing this, myself included, does it because this is what we love to do."

Potter's tape labels have certainly been labors of love. Since 2015, he's run three different imprints: Null Zone, which includes his own music

Some of the "currently listening" tape stacks of cassette label owner Michael Potter. (Photo by Michael Potter)

as well as psychedelic rock from around the world; Garden Portal, dedicated to avant-garde folk; and Serrated Tapes, which releases primarily harsh noise. Growing up in the mid-1990s, Potter made mixtapes off the radio, using his love of music as an escape. "I lived next to large patches of forest," he says. "And I would walk around in the woods to hide from my parents while I smoked weed and listened to my mixtapes on a Walkman." His collection has grown to the point where he has closets full of tapes, as well as a "currently listening" pile that numbers over 500. Potter makes every copy of his label's releases himself, rather than having his titles professionally manufactured. He can usually dub about ten to twenty per day in the downtime he has when he's not at his job. "My life at any given time could literally be just working and dubbing cassettes," he says. "It would probably be easier in the long run to just pay a little extra and have them professionally dubbed, but where's the DIY in that?"

Potter's workload may seem extreme, but he's not the only one making so many tapes. Just ask Mike Haley, who cohosts a podcast and website devoted to cassettes, *Tabs Out*, with his friend Dave Doyen. "We get a *lot* of tapes," he says. "At first we would play stuff from our collections and shit we bought at shows, every once in a while playing something we got in the mail. Now we get about 2,000 tapes a year." "Haley gives me a bag of about 100 tapes every month," adds Doyen. "I take them home and try to keep everything separated in piles of stuff I have listened to and haven't listened to, but at some point, they get all mixed up. When that pile gets too big, the stuff I haven't pulled to play on the show goes back to Mike and he sends them out to writers and other contributors."

The *Tabs Out* podcast is recorded at Haley's Connecticut home, in a room with, as he puts it, "tapes covering almost every spot on the wall." Once a month, the pair pick a stack to play, then talk in between tracks about cassettes or whatever else is on their mind. Titles that don't make it onto the program often get reviewed on the *Tabs Out* website by Haley, Doyen, or their cassette-obsessed comrades. What started as a chance to talk about music with pals has become Haley's way to stay in touch with music. "I don't know if I could keep up with all the labels and projects otherwise," he says. "I'm not sure how anyone does that. I was actually talking with my wife about how weird my music consumption is. I don't really buy anything anymore; I just listen to random stuff that strangers send me. It's bizarre."

Haley claims to listen to at least some portion of every tape he receives, meaning that sometimes dinners at home are accompanied by cassettes playing, "to varying degrees of pleasure within the family," he adds. He keeps a spreadsheet with info about every tape, adding a rating between one and ten so he can rank them in the *Tabs Out* Top 200 at year's end. He and Doyen have yet to tire of cassettes. "They just lend themselves to creative and personalized aesthetics," says Doyen. "It's hard to make a CD-R not remind you of Staples or OfficeMax. It's easy to make a tape look awesome, because before you even do anything to them, they're already awesome little machines that come in a million different shell colors with a million textured patterns on them. Plus, they are in constant danger of being eaten, so they are the bravest of all formats."

One writer *Tabs Out* enlisted to help manage the onslaught of cassettes is Ryan Masteller. When he answered Haley and Doyen's call for scribes in 2017, he was already reviewing cassettes for the blog *Cassette*

Gods at a rather fanatical pace. "At first, I was tasked to write about one tape a week," he remembers. "I quickly eclipsed that quota, and soon enough I had two, three, four posts per week there." In just five years, Masteller posted 830 reviews on *Cassette Gods*, roughly one every other day. Like many of those who run labels and send him music for review, Masteller grew up on tapes, filling up blanks with favorites. Though he veered to other formats when cassettes declined, returning to tapes was like getting back on a bike. "They're so tactile, and they require such a physical act to get them to work," he enthuses. "The process is almost as fascinating as the listening experience itself. . . . You get to see actual magnetic tape spool across a head that reads the music on the tape and plays it back for you. You can watch a physical manifestation of time as it passes."

For Masteller, the underground aspect of current cassette culture only adds to the allure. "Maybe cassettes are the outsider collector's secret," he muses. "The currency of an underground nerd culture whose lexicon is so dense and whose interconnectedness is so dynamic that to infiltrate it is to become part of a tradition so rich in self-reference and so cutting-edge in execution that you feel like you're always on the front line of the most interesting artistic expression. You get to experience the best stuff first, and you get to tell others about it."

Of course, many tape labels want people to hear their music, not leave it underground. Bandcamp has helped in that regard, but so have brick-and-mortar stores that still sell tapes. Some even specialize in it. In the Los Angeles neighborhood of Atwater Village, Jacknife Records and Tapes gives as much space to cassettes as any other format. Trevor Baade bought the shop in 2012, converting it from mostly vinyl stock to equal amounts of LPs, CDs, and cassettes. Baade loves all these formats, but at Jacknife—whose business-card logo is shaped like a tape—cassettes catch your eye first.

"People walking by will think, 'A tape store?'" he explains. "They don't realize there are also records inside. I figure if they come in, they'll realize that it's a record store, but they need to be surprised by the tapes." Even tape collectors themselves might be surprised. In the relatively small, narrow space at Jacknife, the first thing you see as you enter is a custom-made wall of cassette shelves stretching from near the floor up almost to the ceiling. New titles sit next to vintage picks from every imaginable genre: rock, hip-hop, metal, new age, and more. There are

The shelves of cassette tapes that greet
customers when they walk into Los
Angeles' Jacknife Records and Tapes.
(Photo by Marc Masters)

stacks of blank tapes available for anyone wanting to make a mix, as well
as a selection of decks and players, which Jacknife can service if, as often
happens, gear wears out or needs new parts.

"Over the years, I've amassed a lot of stuff," Baade says. "We're really
ground zero for tapes in LA. I would be shocked if there's another place
that has as many tapes as we have." He builds this inventory through a
variety of acquisitions. In the mid-2000s he bought an entire tape distributor that was going out of business, netting him tens of thousands
of titles. Jacknife prices are above thrift-store rate, reflecting Baade's
respect for the format. But they are also reasonable enough that someone could start a collection quickly via purchases at Jacknife or get
decently compensated for tapes they want to part with. "For us, all formats are equally valid, and because of that we don't think of tapes as
leftover junk," he says. "We pay real money—money that people wouldn't
imagine they can get for tapes."

Baade's respect for cassette tapes comes from personal affection, the kind you see in the people who run the labels he carries. "There are so many things I like about them," he says. "The size, the compactness, the fact that you can't jump around from song to song. I like that if I leave a tape running, I'm forced to stay in that zone and listen the way it was intended, in the sequence and length. You can decide to stop it and move on, but if you let it play out, you absorb it in a different way. And I like that they wear out. I like that they're kind of fragile."

Jacknife is one of the few stores anywhere that could fairly call itself a "cassette store," though even Baade sells other formats too. But that dearth of tape-centered outlets didn't stop a group of labels from following the lead of Record Store Day by creating Cassette Store Day. The annual event was founded in 2013 to highlight tape releases and inspire labels to produce titles exclusive to that day. "Unlike Record Store Day," cofounder Jen Long of the UK tape label Kissability told *NME*, "this is less about supporting shops and more about celebrating the cassette format that has been making a comeback for a while." At first, Cassette Store Day caught on, going from fifty exclusive titles in its inaugural year to 300 the next. High-profile bands such as the Flaming Lips and the White Stripes made cassettes for the event, and by 2016, original participants in the UK and the United States were joined by stores in Australia, New Zealand, Germany, France, and Japan. But eventually obstacles arose. In 2019, shortages of ferric oxide needed to make tape slowed down production of exclusives. A year later, partnering label Burger Records was revealed to have a culture of sexual misconduct among its staff and artists, forcing the event to dissolve. The mantle has since been taken up by online cassette merchant Tapehead City, which started Cassette Week in 2021 with a handful of exclusive titles.

The success or failure of such an event is unlikely to change the fact that new, used, and blank tapes continue to be sold in stores. Such establishments might be niche, but Baade sees them as a chance to inform and create new aficionados or resurrect ones who had left their collections for dead. "The number one comment that I get here is when someone walks in and looks at the wall of tapes in amazement," he explains, "and they turn around and tell me, 'I just threw away all my tapes!'"

I love being able to look at a stack of tapes and just put something on," says Doug Kaplan of Hausu Mountain, "versus going through a streaming service and thinking, 'What do I type in? What do I do?' It's so much easier to put on music when there are forty tapes in front of you to choose from." Stacking up the well over 100 tapes released by Hausu Mountain might make your choice of what to play a lot tougher. In the imprint's decade of existence, Kaplan and his cofounder, Maxwell Allison, have tapped into a wealth of uncompromising sounds that put them somewhere near the center of the international cassette underground. Their connections to other labels include trading tapes, sharing artists, and even giving them their own music. "We can just start rattling off names of people who are our close friends," says Kaplan. "Orange Milk, Patient Sounds, NNA, Northern Spy, Constellation Tatsu." "All of us cross streams," says Allison. "There's a huge Venn diagram, just constantly expanding."

In a way emblematic of the current cassette underground, Kaplan and Allison have woven together a sonic and visual aesthetic that unites all their releases while allowing freedom to stretch out and add new sights and sounds as they go. "It's about finding a signature that makes the existence of the label justified," says Allison. "If it's a total grab bag and there are no coherent elements, then it feels arbitrary." "It's important to have a personality so people can genuinely connect to us," adds Kaplan. "The internet is this vast well of endless things you've never heard of, and people are looking for what to trust. If a label has a strong personality, it's easier for people to latch directly on to them."

One of the strongest aspects of Hausu Mountain's personality is presentation. Allison, who designs most of the art, is self-trained on computer software and admits he can't even draw or paint. "I think this forced me to develop an idiosyncratic approach to visual art," he says. "I have to make up for my deficiencies by getting weirder and more focused in other areas." Allison's process includes layering anime frames, comic book panels, screen shots from video games, and even copies of a set of tarot cards he found in a free bin. "My goal then is to twist those seemingly random sources into something that seems coherent, or at least pleasantly incoherent," he explains. "I would say what unites almost all of our art is surreality, otherworldliness, and psychedelia. Gaudy and vivid colors, absurd or grotesque subject matter, LSD-tinged melting and bubbling textures, monstrous eyes, and tentacles."

That's a fair description of the music on Hausu Mountain. Angel Marcloid's Fire-Toolz grafts together digital debris and AM radio pop music. Duo Kill Alters make hyperkinetic workouts that explore childhood traumas. One-man sound blender RXM Reality throws all kinds

Good Willsmith's 2015 cassette tape *Snake Person Generation* on Hausu Mountain, designed by band member and label co-owner Maxwell Allison.
(Photo by Doug Kaplan)

of dissonance and melody into a head-spinning mix. Kaplan's and Allison's own music—both solo and as part of the trio Good Willsmith with Natalie Chami—fits right in with their label's morphing aesthetic. "Many of our artists use interlaced systems of analog and digital gear," says Allison. "Similarly, my art combines harsh or grainy textures from scanned physical objects with glossy, more hi-res imagery from digital sources. The imagery can enliven the music in a new way, and the music can spark connections to elements of the art that seem to defy sense or logic." In addition, the kind of music Hausu Mountain supports takes full advantage of the idiosyncrasies of cassette. "Tapes that have been made in the last ten years are the best analog format for [this kind of] electronic music," says Kaplan. "They have way better bass response and more shimmery treble than an LP, just because of physical limitations of plastic. For a lot of the music that we've released that's incredibly detailed, there is no better-sounding format than a professionally duplicated cassette."

Get Kaplan and Allison talking about what they love about cassettes, and they can go on forever.

> ALLISON: I like supporting artists directly. I like having something that I can touch and look at. I love looking at the artwork. I like having something that I can collect. We're both big collectors.

> KAPLAN: I love when you open up a tape and there's a J-card that has like a hundred panels. Or you pull it out and it has every lyric, and the print is tiny. I also love that a tape is a perfectly pocket-sized item. It's just such a mobile thing, compared to, say, holding a record and walking around town.

> ALLISON: It's like having a very compact rectangular talisman. It's like a little ship in a bottle. It contains such detail and such compression of ideas and imagery. It's almost like you have to squint at it like a jewel to interact with it.

> KAPLAN: Records are big, CDs are super fragile. Cassettes are kind of just right.

> ALLISON: I like when you're playing a tape and you're interrupted and you go do something else and you come back, and it's like the tape has frozen in time and its scroll is still right where you left it.

> KAPLAN: We have an auto-reverse tape deck, and there have been several times where we put on a tape and let it auto-flip for like six hours. It's amazing.

> ALLISON: We don't have to move or do anything.

Since it first got attention in the late 2000s, the cassette comeback has been discussed so often that the question of whether the format is here to stay seems to have an obvious answer. If people have talked about cassettes coming back for over a decade, they must be back. That all depends, of course, on what "back" means. "Back" in terms of being produced by major labels and purchased and played by nonaficionados? It's tough to say if cassettes have ever been back in that way, and if so, it certainly could have been a temporary fad. "Back" in terms of resonating with fans who value their look, sound, feel, price, convenience, and more? That feels more like it, and more like something sustainable.

"It's pretty inconceivable that the circulation of cassettes would ever become a fully mainstream activity again," says Scott Scholz of Tymbal Tapes. "But for folks who share certain kinds of musical proclivities and cultural values, they're finding one another now, and I suspect the broader musical scenes around cassettes will continue to be supportive places no matter what happens to the physical format in the future." Adam Svenson of Eiderdown puts it more bluntly: "Tapes inspire love and loathing. Because of this, tapes are still punk, regardless of the rise in popularity." "We control the distribution of our art, and we make cassettes however we damn well please," says Dylan McConnell of Field Hymns. "You don't need a pressing plant to make tapes, and you can very easily get that tape to fans all over the world via PayPal and the internet. If the art fails, fuck it—you only made seventy-five of them, and in 100 years, they will disappear into the ocean anyway."

This freedom to create, fail, and try again is why cassettes hold such promise and power for independent labels and artists, and why many cassettes made today are as powerful and intriguing as those made decades ago. "It's when you dig into the short runs, the sneaky releases from Bulgaria and Chile, from Stuttgart and Cleveland, that you find the gems, the secrets, the treasures," says Ryan Masteller. "Those are the ones that matter. They're not out there for everyone, and they're not out there to break some sort of new scene or sound. They're out there for themselves, and to find them and to approach them on their own wavelength is the real benefit. These aren't artists and labels who can make large manufacturing commitments, but they do it anyway. And there's an incredible genuineness to that."

On Saturday, March 6, 2021, Lou Ottens passed away at the age of ninety-four. His death made international news, covered by NPR, CNN, the BBC, and many more. "It may be hard to appreciate how radically Lou Ottens changed the audio world," wrote Neil Genzlinger in the *New York Times.* Obituaries were filled with tributes to the cassette tape—to memories of buying blanks, dubbing albums, making mixtapes, recording music, and much more. But for Ottens, looking back had always paled in comparison to moving forward. After all, the man who spearheaded the compact cassette also helped develop a technology that would topple it from its peak: the CD. "The cassette is history," Ottens told *Time* in 2013. "I like it when something new comes." And as he said in *Cassette: A Documentary Mixtape*, "When your time has gone, it's time to disappear. Is there a better product than cassettes? Well, then you stop [making them]."

Fortunately, that's not the way it has worked out. There might be "better" products than cassettes on many levels: sound quality, convenience, reputation, collectability. But that hasn't stopped people from using the cassette tape for everything that makes it great: affordability, durability, portability, personalizability. It all adds up to something that people who discover it tend to get attached to. That's why preserving the cassette tape isn't about something so simple as nostalgia. It's an attempt to protect something that will always have value. Records get scratched, CDs degrade, digital files get lost on dying hard drives—and of course, tapes snag in players and melt inside their cases. But the compact cassette has an uncanny ability to rise from its grave time and again.

Despite his insistence on always looking ahead, Lou Ottens understood all this. His goal of making a device that could fit in people's pockets wasn't just about convenience. It was about giving them a musical format that they could hold and feel, that they could incorporate into their lives, that they could use and copy and modify to make into something their own. Some may marvel that anyone could ever care that much about a small plastic box, but the connections the cassette tape makes can't be severed by advances in technology or even the passage of time. "The people who use it nowadays . . . they love the cassette," Ottens said in *Cassette: A Documentary Mixtape*. "It's not really rational, hmm? It's a rather irrational activity. I like that."

ACKNOWLEDGMENTS

Thanks to everyone who agreed to speak to me and share their knowledge for this book. In alphabetical order: Maxwell Allison, Michael Anderson, Trevor Baade, Lou Barlow, Jed Bindeman, Alan Bishop, Sindre Bjerga, Gene Booth, Britt Brown, Dennis Callaci, Daniel Castrejón, Stuart Chalmers, Fred Church, Liz Clayton, Paul Condon, Tony Coulter, Nate Cross, Davey D, Erin Margaret Day, Amy Denio, Frans de Waard, Reynor Diego, Aaron Dilloway, Toby Dodds, Dave Doyen, Jeff Economy, Kathy Fennessy, John Foster, Jason Gercyz, Mark Gergis, Pete Gershon, Sarah Grady, Seth Graham, Mike Haley, Eamon Hamill, Eric Hardiman, Jonathan Herweg, Nico Hobson, Menghsin Horng, Aadam Jacobs, Robin James, GX Jupitter-Larsen, Doug Kaplan, Chris Kirkley, Farbod Kokabi, Steven Krakow, Willy Leenders, David Lemieux, Ryan Martin, Ryan Masteller, Hal McGee, James McNew, Albert Mudrian, Joe Murray, Paul Nixon, Darryl Norsen, Robert O'Haire, Aki Onda, Henry Owings, Bruce Pavitt, Dwight Pavlovic, Liz Pavlovic, Michael Potter, Jim Powell, Bobby Power, Jack Rabid, Steven Ramsay, Lee Ranaldo, Keith Rankin, David Rauh, Tony Rettman, Mark Rodriguez, Aaron Schnore, Pieter Schoolwerth, Paul Scotton, Brian Shimkovitz, Leslie Singer, Brian Slagel, Howard Stelzer, Joshua Tabbia, Jeff Tartakov, Randall Taylor, Pat Thomas, Robert Wagner, Brian Weitz, and Jason Zeh.

Thanks to those who helped me with research, contacts, feedback, background, history, and many other kinds of information and advice: Jed Bindeman, John Darnielle, Eric Harvey, John Howard, Jesse Jarnow, Calvin Johnson, Kim Kelly, Steve Kiviat, Jeff Krulik, Catherine Lewis, Mark Lore, Bret McCabe, Dylan McConnell, Sommer McCoy, Joseph Nechvatal, Darryl Norsen, Bryan Parker, Nate Patrin, Mark Richardson, Christopher Sienko, Andrew Simon, Howard Stelzer, Gregg Turkington, Melisza Valette, Lance Scott Walker, and Jonathan Williger. Special thanks to Zack Taylor for helping kick-start my research and for graciously giving me access to his interviews with Lou Ottens.

Thanks to all my friends I traded tapes with from childhood to college: Nick Aynsley, Dan Bain, Chrissy Baucom, Laura Boutwell, Walter Carlton, Matthew Dawson, Eric Didul, Lisa Dixon, Mike Gettings, Mike Halverson, Eric Hardiman, Eric Highter, Chris Kelly, Craig Kilcourse, Leigh Tillman Partington, Steve Scafidi, Karen Schoemer, Art Stukas, Bill Tipper, James Wilkins, and Scott Williams. Special thanks to Glen Springer for giving me a copy of the *Toxic Tunes* mixtape when we were first-year students at the College of William and Mary.

Many great thanks to Lucas Church at UNC Press, for suggesting this project, trusting me to write it, answering my questions, tracking down sources, and being a great editor. And many great thanks to Bobby Power for recommending me to Lucas, which made it possible for me to write this book.

Infinite love to my family for supporting me during the writing of this book: my sons Max and Miles, my brothers John and Mike, my mother-in-law Kay Miller, and my wonderful wife who has been so encouraging of my work and accommodating of my challenges, Angela Miller.

BIBLIOGRAPHY

Introduction

Diamond, Michael, and Adam Horovitz. *Beastie Boys Book*. New York: Spiegel and Grau, 2018.

Grady, Sarah. "A Letter Is a Vote for the Future." *Made You a Tape* (blog). November 22, 2018. www.madeyouatape.com/a-letter-is-a-vote-for-the -future.

Hegarty, Paul. "The Hallucinatory Life of Tape." *Culture Machine* 9 (2007). https://culturemachine.net/recordings/the-hallucinatory-life-of-tape.

Taylor, Zack, dir. *Cassette: A Documentary Mixtape*. New York: Seagull and Birch, 2016.

Chapter One

Alexander, Ron. "Stereo-to-Go—and Only You Can Hear It." *New York Times*, July 7, 1980.

Andriessen, Willem. "THE WINNER: Compact Cassette; A Commercial and Technical Look Back at the Greatest Success Story in the History of AUDIO up to Now." *Journal of Magnetism and Magnetic Materials* 193, no. 1 (March 1999): 1–16.

Angus, Robert, and Norman Eisenberg. "Are Cassettes Here to Stay?" *High Fidelity*, July 1969.

Billboard. "'Audio Legend' Envisions Finer Home Equipment." December 11, 1971.

———. "British Govt. Urges New Look at Taping License." November 14, 1981.

———. "Columbia Gains Temporary Ban on Duplication." July 23, 1966.

———. "Europe Grapples with Home Solution." October 27, 1979.

———. "Lear Is Developing Fast-Forward Unit." July 23, 1966.

———. "Standardization Key to Cassette Future." April 8, 1967.

———. "U.K. K-tel Uses Anti–Home Tape Slogan First." November 14, 1981.

———. "U.K. Press Hits WEA Exec's Taping Letter." October 2, 1982.

Bloom, Allan. *The Closing of the American Mind*. Harmondsworth, UK: Penguin, 1989. As cited in Du Gay et al., *Doing Cultural Studies*.

Bottomley, Andrew J. "'Home Taping Is Killing Music': The Recording Industries' 1980s Anti–Home Taping Campaigns and Struggles over Production, Labor and Creativity." *Creative Industries Journal* 8, no. 2 (2015): 123–45.

Bull, Michael. "Investigating the Culture of Mobile Listening: From Walkman to iPod." In *Consuming Music Together: Social and Collaborative Aspects of Music Consumption Technologies*, edited by Kenton O'Hara and Barry Brown, 131–50. Dordrecht, Netherlands: Springer, 2006.

Business Week. "Music Maker for the Masses." February 24, 1968.

Button, S. "Tuning into Rappers." *Money*, July 1981.

Coalition to Save America's Music. "Home Taping to Us . . . Is Like Shoplifting to You." Advertisement. *Billboard*, March 15, 1986.

Chambers, Iain. "A Miniature History of the Walkman." *New Formations*, Summer 1990.

Cusic, Don, Gregory K. Faulk, and Robert P. Lambert. "Technology and Music Piracy: Has the Recording Industry Lost Sales?" *Studies in Popular Culture* 28, no. 1 (2005): 15–24.

Daniel, Eric D., C. Dennis Mee, and Mark H. Clark, eds. *Magnetic Recording: The First 100 Years*. New York: IEEE Press, 1999.

Dormon, Bob. "Are You for Reel? How the Compact Cassette Struck a Chord for Millions." *Register*, August 30, 2013. www.theregister.com/2013/08/30/50_years_of_the_compact_cassette.

Du Gay, Paul, Stuart Hall, Linda Janes, Anders Koed Madsen, Hugh Mackay, and Keith Negus. *Doing Cultural Studies: The Story of the Sony Walkman*. 2nd ed. London: Sage, 2013.

Fantel, Hans. "An Era Ends as Cassettes Surpass Disks in Popularity." *New York Times*, November 21, 1982.

———. "Cassettes—Check before Using." *New York Times*, October 5, 1969.

———. "Home Taping: The Legal Issue Comes to a Boil." *New York Times*, August 29, 1982, section 2.

Foti, Laura. "Home Taping Issues Probed in EIA Study." *Billboard*, October 2, 1982.

Gold, Gerald. "Cassette Sales Booming amid 'Pirating' Dispute." *New York Times*, August 26, 1982.

Gortikov, Stanley M. "Home Taping: Copyright Killer." *Billboard*, May 15, 1982.

Harrington, Richard. "The Record Industry Goes to War on Home Taping." *Washington Post*, June 15, 1980.

——. "Record Rentals: Cashing In on Home Taping." *Washington Post*, June 28, 1981.

Hennessy, Mike. "U.K. Biz in Arms over Taping Levy." *Billboard*, November 14, 1987.

Heylin, Clinton. *Bootleg: The Secret History of the Other Recording Industry*. New York: St. Martin's, 1996.

Holland, Bill. "Home Taping Battle Set to Resume." *Billboard*, November 24, 1984.

Horowitz, Is. "Home Taping Monster Provokes Variety of Reactions." *Billboard*, May 24, 1980.

Hosokawa, Shuhei. "The Walkman Effect." *Popular Music* 4 (1984): 165–80.

Hsu, Hua. "Thanks for the Memorex." *Artforum International*, February 2011.

Jackson, Vincent. "Menace to Society." *Touch* 42, November 1994. As cited in Du Gay et al., *Doing Cultural Studies*.

James, Robin, ed. *Cassette Mythos*. Brooklyn, NY: Autonomedia, 1992.

Jones, Steve. "The Cassette Underground." *Popular Music and Society* 14, no. 1 (1980): 75–84.

——. *Rock Formation: Music, Technology, and Mass Communication*. London: Sage, 1992.

Joseph, Raymond A. "Hey, Man! New Cassette Player Outclasses Street People's Box." *Wall Street Journal*, June 23, 1980.

Keller, Daniel. "Bruce Springsteen's 'Nebraska': A PortaStudio, Two SM57's, and Inspiration." Tascam (website), July 25, 2007. https://tascam.com/us/support/news/481.

Kelley, Frannie. "A Eulogy for the Boombox." *NPR*, April 22, 2009.

Kozak, Roman. "CBS Develops Taping 'Spoiler.'" *Billboard*, October 2, 1981.

Leenders, Willy. *A Hystory of the Future: Into the 20th Century with Hasselt and Philips*. Hasselt, Belgium: Stellingwerff-Waerdenhof Municipal Museum, 1999.

Lichtman, Irv. "A U.S. First: Mango Trying One-Plus-One Island Tapes." *Billboard*, November 7, 1981.

Marsh, Dave. *Glory Days: Bruce Springsteen in the 1980s*. New York: Pantheon, 1987.

McCormick, Moira. "RIAA Chief Hits Home Taping." *Billboard*, January 22, 1983.

McCullaugh, Jim. "TEAC Bows Mini Portastudio." *Billboard*, September 22, 1979.

Millard, Andre. *America on Record: A History of Recorded Sound*. Cambridge: Cambridge University Press, 1995.

Morita, Akio. *Made in Japan*. New York: E. P. Dutton, 1986.

Morton, David. *Off the Record: The Technology and Culture of Sound Recording in America*. New Brunswick, NJ: Rutgers University Press, 2000.

Nathan, John. *Sony: The Private Life*. New York: Houghton Mifflin, 1999.

Newsweek. "Cassettes Are Rolling." April 28, 1969.

New Yorker. "The Walkman." September 21, 1981.

———. "Walkman." January 2, 1989.

NPR Music. "The History of the Boombox." YouTube video, 10:44. April 22, 2009. www.youtube.com/watch?v=e84hf5aUmNA.

Oatman-Stanford, Hunter. "How Boomboxes Got So Badass." *Collectors Weekly*, December 16, 2013. www.collectorsweekly.com/articles/how -boomboxes-got-so-badass.

Ottens, Lou. "Compact Cassette Supremo Lou Ottens Talks to El Reg." Interview by Bob Dormon. *Register*, September 2, 2013, www .theregister.com/2013/09/02/compact_cassette_supremo_lou _ottens_talks_to_el_reg.

———. "The Compact-Cassette System for Audio Tape Recordings." Paper presented at the 31st Convention of the Audio Engineering Society, New York, NY, October 10–14, 1966.

Owerko, Lyle. *The Boombox Project*. New York: Abrams, 2010.

Pareles, Jon. "Record-It-Yourself Music on Cassette." *New York Times*, May 11, 1987.

Robertshaw, Nick. "Artists Pitch Anti–Home Tape Drive in U.K." *Billboard*, November 7, 1981.

Russell, Luther. "Four Tracks and an Attitude." *Tape Op*, January/February 2011.

Sanderson, Susan, and Mustafa Uzumeri. "Managing Product Families: The Case of the Sony Walkman." *Research Policy*, no. 24 (1995): 761–82.

Schoenherr, Steven. "The History of Magnetic Recording." Paper presented at IEEE Magnetics Society Seminar, San Diego, CA, November 5, 2002.

Schönhammer, Rainer. "The Walkman and the Primary World of the Senses." *Phenomenology and Pedagogy* 7, no. 1 (1989): 127–44.

Sills, Beverly. "For Taxes to Offset Loss from Tape." *New York Times*, March 1, 1984.

Sisario, Ben. "When the Beat Came in a Box." *New York Times*, October 15, 2010.

Sony. "Promoting Compact Cassettes Worldwide." Accessed February 3, 2023. www.sony.com/en/SonyInfo/CorporateInfo/History /SonyHistory/2-05.html.

Sugar, Alan. *What You See Is What You Get: My Autobiography*. London: Macmillan, 2010.

Sutherland, Sam. "Taping Losses Near $3 Billion?" *Billboard*, April 3, 1982.

Taylor, Zack, dir. *Cassette: A Documentary Mixtape*. New York: Seagull and Birch, 2016.

Traiman, Stephen. "Home Taping Controversy Clouds Horizon." *Billboard*, August 26, 1978.

Tuhus-Dubrow, Rebecca. *Personal Stereo*. New York: Bloomsbury, 2017.

Van der Lely, P., and G. Missriegler. "Audio Tape Cassettes." *Philips Technical Review* 31, no. 3 (1970): 77–92.

Weber, Heike. "Taking Your Favorite Sound Along: Portable Audio Technologies for Mobile Music Listening." In *Sound Souvenirs*, edited by Karin Bijsterveld and Jose Van Dijc, 69–82. Amsterdam, Netherlands: Amsterdam University Press, 2009.

White, Adam. "British Striving to Apply Brakes to Home Tapings." *Billboard*, August 13, 1977.

———. "Is 'License' Answer to Home Taping?" *Billboard*, August 20, 1977.

Worldwide Independent Inventors Association. "Compact Cassette, History, Invention." November 22, 2009. https://worldwideinvention.com /compact-cassette-history-invention.

Wyman, Jack. "Home Taping: Scapegoat." *Billboard*, July 10, 1982.

Zaleski, Annie. "35 Years Ago: The U.K. Launches the 'Home Taping Is Killing Music' Campaign." *Diffuser*, October 25, 2016.

Chapter Two

Allah, Bilal. "DJ Screw: Givin' It to Ya Slow." *Rap Pages*, July 1995.

Arnold, Jacob. "Ron Hardy at the Music Box." *Red Bull Music Academy*, May 18, 2015. https://daily.redbullmusicacademy.com/2015/05/ron -hardy-at-the-music-box.

Azerrad, Michael. *Our Band Could Be Your Life: Scenes from the American Indie Underground 1981–1991*. New York: Back Bay Books, 2001.

Ball, Jared. *I Mix What I Like: A Mixtape Manifesto*. Oakland, CA: AK Press, 2011.

Baranauskas, Liam. "Paradise." *Oxford American*, no. 107 (Winter 2019). https://oxfordamerican.org/magazine/issue-107/paradise.

Barlow, Lou. "Episode 614: Lou Barlow." Interview by Vish Khanna. *Kreative Kontrol* (podcast), May 25, 2021. http://vishkhanna.com /2021/05/25/ep-614-lou-barlow.

Baumgarten, Mark. *Love Rock Revolution: K Records and the Rise of Independent Music*. Seattle: Sasquatch Books, 2012.

Bessman, Jim. "ROIR Brings Its Punk-Era Rarities to CD." *Billboard*, January 30, 1999.

Brucie B. "The World Famous Brucie Bee of the Legendary Rooftop." Interview by Troy L. Smith. *Foundation*, Fall 2006. http://thafoundation .com/brucie.htm.

Chang, Jeff. *Can't Stop Won't Stop: A History of the Hip-Hop Generation.* New York: St. Martin's, 2005.

Charleston City Paper. "Artist and Musician Charlie McAlister Remembered by Shephard Fairey and Friends." March 2, 2018.

Cills, Hazel. "A Conversation with Liz Phair on *Guyville*, Growing Up, and Getting a Woman's Life into History." *Jezebel*, April 30, 2018. https://jezebel.com/a-conversation-with-liz-phair-on-guyville -growing-up-1825426346.

Curry, Matthew. "R. Stevie Moore: Grandfather of Home Recording?" *Tape Op*, July/August 2011.

DJ Screw. "What's Dirty Down South." *Platinum*, May 2000.

Faison, Justo. "Interview with Justo Faison: The History of Mixtapes." By Dave Cook (@MrDaveyD). Breakdown FM, April 2005. https://soundcloud.com/mrdaveyd/breakdown-fm-intv-w-justo.

——. prod. *Justo Presents: The Mixtape Documentary.* Just Entertainment, 2004. www.youtube.com/watch?v=tJTkNL-lSF8.

Flipside, Hudly. "Beat Happening." *Flipside*, no. 51 (Winter 1986).

Ford, Robert, Jr. "Jive Talking N.Y. DJs Rapping Away in Black Discos." *Billboard*, May 5, 1979.

Godfrey, Sarah. "Scotched Tapes." *Washington City Paper*, November 4, 2005.

Grady, Sarah. "A Letter Is a Vote for the Future." *Made You a Tape* (blog), November 22, 2018. www.madeyouatape.com/a-letter-is-a-vote -for-the-future.

Hancock, Vickie, and Vicki Frudenthal. "Undergraduate Has High Ambitions." *Madison (TN) High School RamPage*, December 31, 1969.

Harrison, Anthony Kwame. "'Cheaper Than a CD, Plus We Really Mean It': Bay Area Underground Hip Hop Tapes as Subcultural Artefacts." *Popular Music* 25, no. 2 (2006): 283–301.

Hopkinson, Natalie. *Go-Go Live: The Musical Life and Death of a Chocolate City.* Durham, NC: Duke University Press, 2012.

Horowitz, Is. "'Illegit Disco Tapes Peddled by Jockeys." *Billboard*, October 12, 1974.

Jardim, Gary. "The Great Facilitator." *Village Voice*, October 2, 1984.

Johnston, Daniel. "Daniel Johnston Interview." By Marta Salicrú. *Time Out Barcelona*, May 17, 2013.

Kerrang! Penpals. June 1983; December 1983.

Klaess, John. *Breaks in the Air: The Birth of Rap Radio in New York City.* Durham, NC: Duke University Press, 2022.

Kool DJ Red Alert. Interview by Monk One. *Red Bull Music Academy*, 2005. www.redbullmusicacademy.com/lectures/dj-red-alert-reel -recognize-reel.

Kranitz, Jerry. *Cassette Culture: Homemade Music and the Creative Spirit in the Pre-internet Age*. Friedrichshafen, Germany: Vinyl-on-Demand, 2020.

Lornell, Kip, and Charles Stevenson Jr. *The Beat! Go-Go Music from Washington, D.C.* Jackson: University Press of Mississippi, 2009.

Masters, Marc. "Sebadoh." *Rockpool*, November 1990.

McConnell, Kathleen F. "The Handmade Tale: Cassette-Tapes, Authorship, and the Privatization of the Pacific Northwest Independent Music Scene." In *The Resisting Muse: Popular Music and Social Protest*, edited by Ian Peddie, 163–76. Canyon: West Texas A&M University, 2006.

Metalcore Fanzine. "Ron Quintana (Interview)." N.d. Accessed February 13, 2023. https://web.archive.org/web/20220626204600/https:// metalcorefanzine.com/rq.html.

Metal Forces. Penbangers. No. 1 (Autumn 1983); no. 2 (Winter 1983–84); no. 21 (1986).

Mills, David. "Go-Go Still Grooving and Moving." *Washington Post*, July 8, 1990.

Mitchell, Gail. "Houston's Hip Hoppin." *Billboard*, October 15, 2005.

MTV News. "Mixtape History." Recovered via Wayback Machine. Date unknown.

Netherton, Jason. *Extremity Retained: Notes from the Death Metal Underground*. London, ON: Handshake, 2014.

O'Brien, Glenn. "At Last: Contortions for the Automobile!" Liner notes to James Chance and the Contortions, *Live in New York*. ROIR, 1981, cassette.

Palmer, Robert. "Cassettes Now Have Material Not Available on Disks." *New York Times*, July 29, 1981.

Pareles, Jon. "Record-It-Yourself Music on Cassette." *New York Times*, May 11, 1987.

Parker, Bryan C. *Beat Happening*. London: Bloomsbury, 2015.

Pavitt, Bruce. *Sub Pop USA: The Subterranean Pop Music Anthology, 1980–1988*. New York: Bazillion Points, 2014.

Pouncey, Edwin. "Adepts of the Craft." *Wire*, November 2008.

Rechler, Glenn. "Home Recording." *New York Press*, May 20, 1988.

Salkind, Micah E. *Do You Remember House? Chicago's Queer of Color Undergrounds*. New York: Oxford University Press, 2019.

Skillz, Mark. "DJ Hollywood: The Original King of New York." *Cuepoint*, November 19, 2014. https://medium.com/cuepoint/dj-hollywood-the -original-king-of-new-york-41b131b966ee.

Slagel, Brian. *For the Sake of Heaviness: The History of Metal Blade Records.* New York: BMG, 2017.

Spencer, Amy. *DIY: The Rise of Lo-fi Culture.* London: Marion Boyars, 2005.

Tape Deck Wreck. "Grandmaster Flash—NO COPIES!" YouTube video, 1:21:41. Posted April 7, 2020. www.youtube.com/watch?v=dG0Y 6eHS_J4.

Thomas, Marshall, Djibril Ndiaye, Maurice Garland, and Tai Saint-Louis. *The Art behind the Tape.* Axiom Blue Corp., 2014.

Toop, David. *The Rap Attack: African Jive to New York Hip-Hop.* Boston: South End, 1984.

Walker, Lance Scott. *DJ Screw: A Life in Slow Revolution.* Austin: University of Texas Press, 2022.

Wall, Mick. *Enter Night: A Biography of Metallica.* New York: St. Martin's, 2010.

———. *Run to the Hills: Iron Maiden, the Authorized Biography.* London: Sanctuary, 2001.

Yazdani, Tarssa. *Hi How Are You? The Definitive Daniel Johnston Handbook.* New York: Soft Skull, 2000.

Chapter Three

Akita, Masami. "The Beauty of Noise: An Interview with Masami Akita of Merzbow." Interview by Chad Hensley. In *Audio Culture*, edited by Christoph Cox and Daniel Warner, 59–64. New York: Continuum, 2004.

Bailey, Thomas Bey William. *Unofficial Release: Self-Released and Handmade Audio in Post-industrial Society.* Lexington, KY: Belsona Books, 2012.

Baroni, Vittore. "Memo from a Networker." *Lomholt Mail Art Archive.* Accessed March 4, 2021. www.lomholtmailartarchive.dk/texts/vittore -baroni-memo-from-a-networker.

Bath, Tristan. "Splice of Life." *Wire*, February 2019.

Campau, Don. "Andy Xport." *Living Archive of Underground Music*, April 27, 2003. http://livingarchive.doncampau.com/interviews/andy-xport.

———. "A Brief History of Cassette Culture." *Living Archive of Underground Music*, August 29, 2009. http://livingarchive.doncampau.com/about /a-brief-history-of-cassette-culture.

———. "Dino DiMuro." *Living Archive of Underground Music*, April 27, 2003. http://livingarchive.doncampau.com/interviews/dino-dimuro.

———. "Hal McGee." *Living Archive of Underground Music*, November 17, 2011. http://livingarchive.doncampau.com/interviews/hal-mcgee.

———. "Insane Music." *Living Archive of Underground Music*, September 24, 2011. http://livingarchive.doncampau.com/tape_labels/insane-music.

——. "John Foster." *Living Archive of Underground Music*, November 2, 2011. http://livingarchive.doncampau.com/interviews/john-foster.

——. "Lord Litter." *Living Archive of Underground Music*, November 2, 2011. http://livingarchive.doncampau.com/interviews/lord-litter.

——. "Robin James." *Living Archive of Underground Music*, November 2, 2011. http://livingarchive.doncampau.com/interviews/robin-james.

——. "R. Stevie Moore." *Living Archive of Underground Music*, November 2, 2011. http://livingarchive.doncampau.com/interviews/r-stevie-moore.

——. "Sound of Pig." *Living Archive of Underground Music*, September 24, 2011. http://livingarchive.doncampau.com/tape_labels/sound-of-pig.

——. "Vittore Baroni." *Living Archive of Underground Music*, October 16, 2011. https://livingarchive.doncampau.com/early_experiences/vittore -baroni.

——. "Zan Hoffman." *Living Archive of Underground Music*, November 1, 2011. http://livingarchive.doncampau.com/interviews/zan-hoffman.

Cowley, Julian. "José Maceda: Creative Composition." *Wire*, March 2019.

——. "Natural Cycles: Watch José Maceda's *Cassettes 100* Reimagined by Aki Onda." *Wire*, December 2021. https://www.thewire.co.uk/in-writing /essays/natural-cycles-watch-jose-maceda-s-cassette-100-reimagined-by -aki-onda.

Daniel, Drew. *20 Jazz Funk Greats*. London: Continuum, 2008.

DeRogatis, Jim. *Staring at Sound: The True Story of Oklahoma's Fabulous Flaming Lips*. New York: Three Rivers, 2006.

"Graf Hauen Tapes." Tape-Mag. Accessed May 5, 2021. https://tape-mag .com/Graf_Haufen_Tapes+LABELS-1-1-410-2.html.

Hegarty, Paul. *Noise/Music: A History*. New York: Continuum, 2007.

Heylin, Clinton. *Bootleg: The Secret History of the Other Recording Industry*. New York: St. Martin's, 1996.

Hix, Lisa. "Cassette Revolution: Why 1980s Tape Tech Is Still Making Noise in Our Digital World." *Collectors Weekly*, June 2, 2015. https://www .collectorsweekly.com/articles/cassette-revolution.

James, Robin, ed. *Cassette Mythos*. Brooklyn, NY: Autonomedia, 1992.

Keenan, David. "Burning Chrome." *Wire*, November 2008.

Kennedy, Randy. "While the Artists Chatted, He Taped." *New York Times*, April 28, 2010.

Komurki, John Z. *Cassette Cultures: Past and Present of a Musical Icon*. Salenstein, Switzerland: Benteli, 2019.

Kranitz, Jerry. *Cassette Culture: Homemade Music and the Creative Spirit in the Pre-internet Age*. Friedrichshafen, Germany: Vinyl-on-Demand, 2020.

Margolis, Al. *Cassette Culture: Sound of Pig Music 1984–1990*. New York, NY: 98.6 Press, 2022.

——. "In Conversation with Al Margolis (If, Bwana and Sound of Pig Music)." Interview by Greh Holger, Tara Connelly, and Mike Connelly. *Noisextra* (podcast), March 2, 2022. www.noisextra.com/2022/03/02 /in-conversation-with-al-margolis-if-bwana-and-sound-of-pig-music.

McGee, Hal, ed. *Electronic Cottage*, no. 1 (April 1989); no. 2 (September 1989).

——. "Girls on Fire." HalTapes. www.haltapes.com/girls-on-fire.html.

McGraith, Donal. "Anti-Copyright and Cassette Culture." In *Sound by Artists*, edited by Dan Lander and Micah Lexier, 73–87. Toronto, ON: Art Metropole/Walter Phillips Gallery, 1990.

Milne, Bruce, and Andrew Maine, eds. *Fast Forward*. Archives. http://spill -label.org/FastForward.

Moore, R. Stevie. "Do-It-Yourself till You Bleed." Interview by Alfred Boland. *Perfect Sound Forever*, June 2000. www.furious.com/perfect /rsteviemoore.html.

ND, nos. 4–6, 8, 13, 19 (1982–87).

Nechvatal, Joseph, ed. *MINÓY*. Brooklyn, NY: Punctum, 2014.

Novak, David. *Japanoise: Music at the Edge of Circulation*. Durham, NC: Duke University Press, 2013.

No Vibe (blog). "The Flaming Lips' Boombox Experiments." Accessed November 4, 2020. https://novibe.wordpress.com/2013/04/15/the -flaming-lips-boombox-experiments-jpg (no longer available).

Olson, John. "John Olson: American Tapes Interview." Interview by American Tapes a Day. YouTube video, 1:40:43. Posted September 21, 2020. https://www.youtube.com/watch?v=bEs-F_qu1y8

Onda, Aki. "Cassette Memories." Aki Onda (website). https://akionda.net /Cassette-Memories.

——. Interview by Birkut. *Tiny Mix Tapes*, August 20, 2013. www .tinymixtapes.com/features/aki-onda.

Patterson, Archie. "40 Years On." Liner notes to *American Cassette Culture: Recordings 1971–1983*. Vinyl-on-Demand, 2015, 8 vinyl LPs.

Plunkett, Daniel. "In Conversation with Daniel Plunkett (ND Magazine)." Interview by Greh Holger, Tara Connelly, and Mike Connelly. *Noisextra* (podcast), April 6, 2022. www.noisextra.com/2022/04/06/in -conversation-with-daniel-plunkett-nd-magazine.

Richardson, Mark. Unpublished material provided by the author.

——. *Zaireeka*. London: Continuum, 2010.

Sherwood, Lydia. "Spotlight: R. Stevie Moore." *Goldmine* 10, iss. 2, no. 93, February 1984.

Staub, Ian Matthew. "Redubbing the Underground: Cassette Culture in Transition." Master's thesis, Wesleyan University, 2010.

Sullivan, James. "Lips' 'Experiment' Smacks of Novelty." *SF Gate*, February 24, 1998. www.sfgate.com/entertainment/article/Lips-Experiment -Smacks-of-Novelty-Band-3310553.php.

Tau, Michael. *Extreme Music: From Silence to Noise and Everything in Between*. Port Townsend, WA: Feral House, 2022.

"VEC Audio." Tape-Mag. Accessed January 6, 2023. https://tape-mag.com /VEC_Audio+LABELS-1-1-1032-2.html.

Chapter Four

Billboard. "Court Names 'Evil Genius.'" May 27, 1978.

"Collecting Grateful Dead Tapes: We're Living In The Golden Age." Live Grateful Dead Music. Accessed May 17, 2019. www.live-grateful-dead -music.com/grateful-dead-tapes.html.

Coscarelli, Joe. "'Tapers' at the Grateful Dead Concerts Spread the Audio Sacrament." *New York Times*, July 5, 2015.

Cummings, Alex Sayf. *Democracy of Sound*. New York: Oxford University Press, 2013.

Dwork, John. *Dupree's Diamond News* 1, nos. 1 and 2 (May and June 1987).

Eng, Monica. "Chicago's Hidden Indie Rock Archive." *WBEZ*, September 29, 2019. https://interactive.wbez.org/curiouscity/taping-guy.

Getz, Michael M., and John R. Dwork. *The Deadhead's Taping Compendium*. Vol. 1: *1959–1974*. New York: Holt, 1998.

———. *The Deadhead's Taping Compendium*. Vol. 2: *1975–1985*. New York: Holt, 1999.

———. *The Deadhead's Taping Compendium*. Vol. 3: *1986–1995*. New York: Holt, 2000.

Glover, Tony. "Live at Max's Kansas City." *Rolling Stone*, August 3, 1972.

Hall, Andrew. Liner notes to Can, *Live in Stuttgart 1975*. Spoon/Mute, 2021, 3 vinyl LPs.

Hall, David. "A Mapleson Afterword: Further Notes on the Mapleson Cylinder Project at the Rogers and Hammerstein Archives of Recorded Sound, New York Public Library." *ARSC Journal* 14, no. 1 (1982): 6.

———. "The Mapleson Cylinders: An Historical Introduction." New York Public Library (website). https://web.archive.org/web/20190223072117 /http://digilib.nypl.org/dynaweb/millennium/mapleson/@Generic __BookTextView/309.

Icepetal. "Learning the Ropes to a Forgotten Trade." *Deadlistening* (blog), May 13, 2009. www.deadlistening.com/2009/05/learning-ropes-to -forgotten-trade.html.

Jackson, Blair, and Regan McMahon, eds. *Golden Road*, nos. 1–4 (1984).

Jarnow, Jesse. "Early Tapers, the United Dead Freaks of America, and the Dawn of *Relix*." *Relix*, April 15, 2014.

———. *Heads: A Biography of Psychedelic America*. New York: Da Capo, 2016.

———. "The Invisible Hit Parade: How Unofficial Recordings Have Flowered in the 21st Century." *Wired*, November 21, 2018.

Jones, Peter, and Nick Robertshaw, "British See Gains over Bootleggers." *Billboard*, June 17, 1978.

Kinney, David. *The Dylanologists: Adventures in the Land of Bob*. New York: Simon and Schuster, 2015.

Kippel, Les, ed. *Dead Relix* 1, no. 1 (November/December 1974); 2, no. 6 (November/December 1975).

Lewiston (ME) Daily Sun. "Mass. Student Spins Frisbee into Degree." November 15, 1983.

Lichtenstein, Grace. "Tape 'Bootleggers' Still Active." *New York Times*, June 5, 1971.

Marshall, Lee. *Bootlegging: Romanticism and Copyright in the Music Industry*. London: Sage, 2005.

———. "The Effects of Piracy upon the Music Industry: A Case Study of Bootlegging." *Media, Culture, and Society* 6, no. 2 (2004): 163–81.

———. "For and Against the Record Industry: An Introduction to Bootleg Collectors and Tape Traders." *Popular Music* 22, no. 1 (2003): 57–62.

McNally, Dennis. *Long Strange Trip: The Inside History of the Grateful Dead*. New York: Three Rivers, 2003.

Moore, Cason A. "Tapers in a Jam: Trouble Ahead or Trouble Behind." *Columbia Journal of Law and the Arts* 30, no. 3–4 (2007): 625–54.

Mullen, Shaun D., ed. *In Concert Quarterly* 1, no. 1 (September 1979); 1, no. 2 (January 1980).

Neumann, Mark, and Timothy A. Simpson. "Smuggled Sound: Bootleg Recording and the Pursuit of Popular Memory." *Symbolic Interaction* 20, no. 4 (1997): 319–41.

New Yorker. "Librarian." Talk of the Town. December 18, 1935.

———. "The Mapleson Cylinders." December 9, 1985.

Paumgarten, Nick. "Deadhead: The Afterlife." *New Yorker*, November 26, 2012.

Phillips, Amy. "Animal Collective License First Legal Grateful Dead Sample Ever." *Pitchfork*, July 17, 2009. https://pitchfork.com/news/35966 -animal-collective-license-first-legal-grateful-dead-sample-ever.

Reiff, Corbin. "A Brief History of Live-Concert Bootlegging." *AV Club*, September 23, 2015. www.avclub.com/a-brief-history -of-live-concert-bootlegging-1798284737.

Ressner, Jeffrey. "Bootlegs Go High Tech." *Rolling Stone*, May 30, 1991, 15–16.

Rosen, Charley. "Mr. 'Tapes' of Brooklyn: He Rules the Grateful Dead Tape Empire." *Rolling Stone*, October 11, 1973.

Schneider, Jason. "Bootlegging: The Underground Pipeline Is Live Music's Lifeline." *Exclaim!*, February 1, 2000. https://exclaim.ca/music/article /bootlegging-underground_pipeline_is_live_musics.

Schonberg, Harold C. "Voices of 80 Years Ago Form a Priceless Legacy." *New York Times*, February 9, 1986.

Thompson, Cole. "Hot Wax: The Lionel Mapleson Story." *My Inwood*, January 13, 2016. https://myinwood.net/35328.

Toth, James J. "Boots in Transit: An Appreciation of the Dead on Cassette." *Aquarium Drunkard*, August 13, 2019. https://aquariumdrunkard .com/2019/08/13/boots-in-transit-an-appreciation-of-the-dead -on-cassette.

Unterberger, Richie. *The Velvet Underground Day-by-Day*. London: Jawbone, 2009.

Vettel, Phil. "Bootlegs: Sound May Vary, but Illegality Is Clear." *Chicago Tribune*, June 28, 1986.

Waddell, Ray. "The Dead Still Live for the Road." *Billboard*, July 3, 2004.

Chapter Five

Castelo-Branco, Salwa El-Shawan. "Some Aspects of the Cassette Industry in Egypt." *World of Music* 29, no. 2 (1987): 32–48.

Faber, Tom. "The Keen Collectors Battling to Preserve Arab Music." *Financial Times*, June 7, 2019.

Frame, Charlie. Review of *Obaa Sima* by Ata Kak. *Quietus*, March 26, 2015. https://thequietus.com/articles/17498-ata-kak-obaa-sima-review.

Gergis, Mark. "Memorabilia: Collecting Sounds with . . . Mark Gergis." Radio Web MACBA, September 22, 2011. Audio recording, 41:58. https://rwm.macba.cat/en/extra/memorabilia-collecting-sounds -conversation-mark-gergis-his-sound-collection.

——. "Remembering Syria: Mark Gergis of Sublime Frequencies Interviewed." Interview by John Doran. *Quietus*, October 17, 2013. https://thequietus.com/articles/13623-mark-gergis-interview-sublime -frequencies-dabke-syria.

Jones, Mikey IQ. "The Incredible Story Of Ata Kak's *Obaa Sima*, the Original Awesome Tape from Africa." *Fact Mag*, March 10, 2015. www .factmag.com/2015/03/10/the-incredible-story-of-ata-kaks-obaa-sima -the-original-awesome-tape-from-africa.

Manuel, Peter. *Cassette Culture*. Chicago: University of Chicago Press, 1993.

——. "The Cassette Industry and Popular Music in North India." *Popular Music* 10, no. 2 (1991): 189–204.

Masters, Marc. "Meet the Sun City Girls." In Veal and Kim, *Punk Ethnography*, 55–63.

Simon, Andrew. "Censuring Sounds: Tapes, Taste, and the Creation of Egyptian Culture." *International Journal of Middle East Studies*, no. 51 (2019): 233–56.

——. *Media of the Masses: Cassette Culture in Modern Egypt*. Palo Alto, CA: Stanford University Press, 2022.

Suryadi, Minangkabau. "Commercial Cassettes and the Cultural Impact of the Recording Industry in West Sumatra." *Asian Music* 34, no. 2 (2003): 51–89.

Sutton, R. Anderson. "Commercial Cassette Recordings of Traditional Music in Java: Implications for Performers and Scholars." *World of Music* 27, no. 3 (1985): 23–45.

——. *Traditions of Gamelan Music in Java: Musical Pluralism and Regional Identity*. Cambridge: Cambridge University Press, 1991.

Van Zanten, Wim. "Musical Aspects of Popular Music and Pop Sunda in West Java." In *Sonic Modernities in the Malaly World*, edited by Bart Barendregt, 323–52. Leiden, Netherlands: Brill: 2014.

Veal, Michael E., and E. Tammy Kim. *Punk Ethnography: Artists and Scholars Listen to Sublime Frequencies*. Middletown, CT: Wesleyan University Press, 2016.

Wallach, Jeremy. "Exploring Class, Nation, and Xenocentrism in Indonesian Cassette Retail Outlets." *Indonesia*, no. 74 (2002): 79–102.

Wallis, Roger, and Krister Malm. *Big Sounds from Small Peoples: The Music Industry in Small Countries*. London: Constable, 1984.

Widianto, Stanley. "Yess Records, Indonesia's Influential Bootleg Cassette Label." *Red Bull Music Academy*, February 28, 2017. https://daily .redbullmusicacademy.com/2017/02/yess-records-feature.

Wong, Deborah. "Thai Cassettes and Their Covers: Two Case Histories." *Asian Music* 21, no. 1 (1989–90): 78–104.

Chapter Six

Amberson, Joshua James, ed. The Cassette Tape Issue. *Basic Paper Airplane*, no. 13 (2022).

Bitner, Jason, ed. *Cassette from My Ex: Stories and Soundtracks of Lost Loves*. New York: St. Martin's, 2009.

Browne, David. "Outtakes Calling: Inside Joe Strummer's Personal Archives." *Rolling Stone*, July 27, 2022.

Day, Erin Margaret. "Seasonal Mixtape Series." *Come Away with EMD* (blog). Accessed September 21, 2021. https://comeawaywithemd.com /emds-seasonal-mixtape-series.

———. "Winter 2022//Recorded Syntax." *Come Away with EMD* (blog), March 29, 2022. https://comeawaywithemd.com/2022/03/29/winter -2022-recorded-syntax.

Diamond, Michael, and Adam Horovitz. *Beastie Boys Book*. New York: Spiegel and Grau, 2018.

Fox, Kamal. "Mixed Feelings: Notes on the Romance of the Mixed Tape." *Rhizomes*, no. 5 (Fall 2002). http://rhizomes.net/issue5/fox.html.

Grady, Sarah. "A Letter Is a Vote for the Future." *Made You a Tape* (blog), November 22, 2018. http://www.madeyouatape.com/a-letter-is-a-vote -for-the-future.

———. "All through a Life." *Made You a Tape* (blog), February 4, 2017. http://www.madeyouatape.com/all-through-a-life.

———. "Wouldn't Want to Turn Around and Fake It." *Made You a Tape* (blog), April 8, 2017. http://www.madeyouatape.com/jeff.

———. "You Got Your Good Thing and I've Got Mine." *Made You a Tape* (blog), October 17, 2017. http://www.madeyouatape.com/got-good-thing -ive-got-mine.

Hegarty, Paul. "The Hallucinatory Life of Tape." *Culture Machine* 9 (2007). https://culturemachine.net/recordings/the-hallucinatory-life-of -tape.

Jansen, Bas. "Tape Cassettes and Former Selves: How Mix Tapes Mediate Memories." In *Sound Souvenirs*, edited by Karin Bijsterveld and Jose Van Dijc, 43–54. Amsterdam, Netherlands: Amsterdam University Press, 2009.

Lacasse, Serge, and Andy Bennett. "Mix Tapes, Memory, and Nostalgia: An Introduction to Phonographic Anthologies." In *The Pop Palimpsest: Intertextuality in Recorded Popular Music*, edited by Lori Burns and Serge Lacasse, 313–29. Ann Arbor: University of Michigan Press, 2018.

Moore, Thurston, ed. *Mix Tape: The Art of Cassette Culture*. New York: Universe, 2005.

Paul, James. "Last Night a Mix Tape Saved My Life." *Guardian*, September 25, 2003.

Rollins, Henry. "Confessions of a Tape Trader." *LA Weekly*, March 5, 2015.

Sheffield, Rob. *Love Is a Mix Tape*. New York: Crown, 2007.

Sky News. "Lost Mixtape Washes Up on Beach 25 Years Later—and It Still Works." February 14, 2020.

Sylvester, Nick. "It's Just a Cassette." *Pitchfork*, September 6, 2013. https://pitchfork.com/features/oped/9212-its-just-a-cassette.

Woodyatt, Amy. "Woman Reunited with Mixtape More than 20 Years after She Lost It." *CNN*, February 14, 2020.

Chapter Seven

Amorosi, Ad. "Hit the Deck: The Cassette Tape Revival Is in Full Swing." *Flood Magazine*, July 1, 2021.

Baldwin, Rosencrans. "Our Misplaced Nostalgia for Cassette Tapes." *New York Times*, December 23, 2015.

BBC News. "How the CD Was Developed." August 7, 2007.

———. "Not Long Left for Cassette Tapes." June 17, 2005.

Bershidsky, Leonid. "Cassette Tapes Are Making a Comeback: But It's Not about the Music." *Los Angeles Times*, August 6, 2019.

Boles, Benjamin. "Cassette Comeback: Tapes and Weirdo Music Go Hand in Hand." *Now Toronto*, March 9, 2016.

Bowe, Tucker. "Classic Cassette Tapes Are Making a Comeback." *Gear Patrol*, May 12, 2021.

Brown, August. "A Sonic Rewind." *Los Angeles Times*, August 1, 2010.

Butterworth, Brent. "Don't Call It a Comeback: Cassettes Have Sounded Lousy for Years (and Still Do!)." *New York Times*, October 26, 2021.

Chaney, Jen. "Why Is the Cassette Tape All Over Pop Culture?" *Vulture*, July 21, 2017.

Chappell, Bill. "Lou Ottens, Inventor of the Cassette Tape, Has Died." *NPR*, March 10, 2021.

Charara, Sophie. "The Unlikely Cassette Comeback Isn't Over Yet: Sales Are Up in 2019." *Wired*, July 21, 2019.

Coleman, Mark. "Can Cassette Tapes Be Cool Again?" *CNN*, September 30, 2013.

Dailey, Kate. "Press Rewind: The Cassette Tape Returns." *BBC Montreal*, May 20, 2013.

DeLuca, Dan. "Move Over, Vinyl: Cassette Tapes Are the New Old Thing to Love." *Philadelphia Inquirer*, March 22, 2021.

Demers, Joanna. "Cassette Tape Revival as Creative Anachronism." *Twentieth-Century Music* 14, no. 1 (2017): 109–17.

Detwiler-George, Jacqueline. "Cassette Tapes Are Back, Don't You Dare Call Them Obsolete." *Popular Mechanics*, November 12, 2018.

Dezember, Ryan, and Anne Steele. "A Global Shortage of Magnetic Tape Leaves Cassette Fans Reeling." *Wall Street Journal*, November 3, 2017.

Dowling, Tim. "Could These Old Cassette Tapes Be My Cash in the Attic?" *Guardian*, November 7, 2017.

Eley, Craig. "Technostalgia and the Resurgence of Cassette Culture." In *The Politics of Post-9/11 Music: Sound, Trauma, and the Music Industry in the Time of Terror*, edited by Joseph P. Fisher and Brian Flota, 43–54. New York: Routledge, 2011.

Enis, Eli. "Cassette Me, Please: The Future of Vinyl and How Cassettes May Offer the Answer." *Uproxx*, February 28, 2018. https://uproxx.com /music/cassette-sales-vinyl-indie-labels-streaming.

Gallagher, David F. "For the Mix Tape, a Digital Upgrade and Notoriety." *New York Times*, January 30, 2003.

Genzlinger, Neil. "Lou Ottens, Father of Countless Mixtapes, Is Dead at 94." *New York Times*, March 11, 2021.

Gomez, Camilo. "The Cassette Tape Underground Resurgence." *Issue Number One*, April 12, 2014. https://issuenumberone.journalism.cuny .edu/2014/04/12/the-cassette-tape-underground-resurgence.

Grossman, David. "We're Facing a Worldwide Cassette Shortage, so Hug Your Nearest Hipster." *Popular Mechanics*, October 11, 2019.

Hancock, Amanda. "Cassette Tapes Make Their Comeback, as Seen in the Colorado Springs Music Scene." *Colorado Springs (CO) Gazette*, April 26, 2021.

Hemness, Taylor. "Missouri Company at the Center of Cassette Tape Resurgence." *KSHB*, February 9, 2018. www.kshb.com/news/science -tech/return-of-the-cassette-tape.

Hogan, Marc. "This Is Not a Mixtape." *Pitchfork*, February 22, 2010. https://pitchfork.com/features/article/7764-this-is-not-a-mixtape.

Homan, Gregory J. "The World Was Running Out of Cassette Tape: Now It's Being Made in Springfield." *Springfield (MO) News-Leader*, January 7, 2018.

Hunt, Michael. "Why the Cassette Tape Is Still Not Dead." *Rolling Stone*, April 18, 2016.

Hunt, Nigel. "Are Cassettes Coming Back Like Vinyl? Doesn't Really Sound Like It." *CBC News*, October 8, 2016.

Iabal, Nosheen. "It's Cool to Spool Again as the Cassette Returns on a Wave of Nostalgia." *Guardian*, February 23, 2019.

Johnson, Christopher, and Chloe Adams. "Dutch Inventor of the Cassette Tape, Lou Ottens, Dies Age 94." *CNN*, March 11, 2021.

Johnson, Lawrence B. "Against All Odds, Good Old Cassettes Lead a Double Life." *New York Times*, June 4, 1995.

King, Mac. "Resurgence of Cassette Tapes Keeps Brooklyn Manufacturer Busy." *Fox 5 New York*, November 10, 2020. www.fox5ny.com/news /resurgence-of-cassette-tapes-keeps-brooklyn-manufacturer-busy.

Kotsikonas, Stephanie. "Audio Cassettes' Unlikely Revival." *Dollars and Sense* (blog), June 1, 2016. https://blogs.baruch.cuny.edu /dollarsandsense/2016/06/01/an-unlikely-audio-cassette-revival.

Kozinn, Alan. "The Future Is Digital." *New York Times*, April 13, 1980.

Kreps, Daniel. "Lou Ottens, Inventor of the Audio Cassette Tape, Dead at 94." *Rolling Stone*, March 10, 2021.

Ledsom, Alex. "Global Revival of Cassettes Has Flourished during the Pandemic." *Forbes*, March 28, 2021.

Long, Jen. "Why We've Created Cassette Store Day (and Why It's Not Just Hipster Nonsense)." *NME*, July 12, 2013. www.nme.com/blogs/nme -blogs/why-weve-created-cassette-store-day-and-why-its-not-just -hipster-nonsense-22948.

Los Angeles Times. "Reelin' in the Years: Cassette Tapes Still Have Their Devotees." August 9, 2007.

Marsh, Calum. "Reconsidering the Revival of Cassette Tape Culture." *Pop Matters*, December 3, 2009. www.popmatters.com/116282 -reconsidering-the-revival-of-cassette-tape-culture-2496127682.htm.

Marshall, Scott. "Long Live the Humble Audio Cassette: A Eulogy." In *Sounding Off! Music as Subversion/Resistance/Revolution*, edited by Ron Sakolsky and Fred Wei-han Ho, 211–14. Brooklyn, NY: Autonomedia.

Merelli, Annalisa. "Audio Cassettes Are Back—because Hipsters." *Quartz*, January 30, 2016. https://qz.com/570954/audio-cassettes-are-back-and -here-to-stay.

Metro UK. "Death of the Cassette Tape." May 7, 2007.

Moore, Joseph. "Retro-Futurism: Are Cassettes Still a Thing?" *Treble*, March 22, 2016. www.treblezine.com/28657-retro-futurism-cassette-resurgence.

Moore, Sam. "Who the Hell Is Buying Cassettes in 2020? NME Investigates." *NME*, July 21, 2020.

Newman, Andrew Adam. "Cassette Tape Going the Way of the Eight-Track." *New York Times*, July 28, 2008.

Olivarez-Giles, Nathan. "Why Cassette Tapes Are Making a Comeback." *Wall Street Journal*, March 9, 2017.

Owsinski, Bobby. "The Real Meaning of Cassette Store Day." *Forbes*, August 5, 2013.

Padmore, Russell. "Cassette Tapes Make a Comeback." *Marketplace*, November 22, 2017.

Pearl, Max. "RIAA Denies Increase in Cassette Sales amid Reports of a Resurgence." *Resident Advisor*, February 22, 2016.

Qasim, Farshad. "The Revival of the Compact Cassette: Meet the Company Running the Last Factory in the World Producing Analogic Audio Tapes." *Soundesign*, September 21, 2015.

Renaldi, Adi. "This Small Duplication Shop Is at the Center of Indonesia's Cassette Revival." *Noisey*, November 19, 2016. www.vice.com/en/article /4xyqk9/this-music-pirate-king-found-a-new-life-in-the-underground -music-scene.

Rogers, Jude. "Total Rewind: 10 Key Moments in the Life of the Cassette." *Guardian*, August 30, 2013.

Rothman, Lily. "Rewound: On Its 50th Birthday, the Cassette Tape Is Still Rolling." *Time*, August 12, 2013.

Roy, Elodie. "Cassette Fever in the Age of Bandcamp." *Pop Matters*, April 22, 2014. www.popmatters.com/180841-cassette-fever-in-the-age-of -bandcamp-2495668918.html.

Schonfeld, Zach. "Is the Cassette Renaissance for Real?" *Newsweek*, September 27, 2014.

Segal, Dave. "Baby, I'm for Reel: Unspooling the Affordable, Accessible Microeconomy of the Cassette Revival." *Stranger*, March 9, 2016.

Sessa, Sam. "Cassettes Make an Unlikely Comeback in Baltimore." *Baltimore Sun*, April 15, 2011.

Sokol, Zach. "Cassette Tapes Are Almost Cool Again." *Noisey*, August 21, 2013. www.vice.com/en/article/rbdk3r/cassette-tapes-are-almost-cool -again.

Tanikawa, Miki. "Starting with Japan, the Compact and Recordable Digital Medium Is Eclipsing Tape: The MiniDisc Takes On the Audio Cassette." *New York Times*, August 6, 1998.

Taylor, Iain. "Audio Cassettes: Despite Being 'a Bit Rubbish,' Sales Have Doubled during the Pandemic—Here's Why." *Conversation*, March 19, 2021.

———. "From Analogue to Digital, from Pragmatism to Symbolism: The Cassette Tape as a Hybrid Artefact in Contemporary Popular Music." Paper presented at the Westmister-Goldsmiths Symposium for Student Research in Popular Music, London, June 24, 2015.

Taylor, Zack, dir. *Cassette: A Documentary Mixtape*. New York: Seagull and Birch, 2016.

Titlow, John Paul. "Music's Weird Cassette Tape Revival Is Paying Off." *Fast Company*, January 12, 2017.

Walker, Rob. "Hitting Rewind on the Cassette Tape." *New York Times*, April 23, 2010.

Williams, Lisa, and Andy Dykes. "Farewell to Cassettes: Tales of the Tape." *Independent*, May 9, 2007.

Williams, Steven. "For Car Cassette Decks, Play Time Is Over." *New York Times*, February 4, 2011.

Zaldua, Chris. "Why the Cassette Resurgence Isn't Going Away." *KQED*, July 12, 2017. www.kqed.org/arts/13634049/why-the-cassette-resurgence -isnt-going-away.

INDEX